FUTURE ROLES OF U.S. NUCLEAR FORCES

IMPLICATIONS FOR U.S. STRATEGY

GLENN C. BUCHAN

DAVID MATONICK

CALVIN SHIPBAUGH

RICHARD MESIC

Prepared for the United States Air Force

Approved for public release; distribution unlimited

RAND

Project AIR FORCE

The research reported here was sponsored by the United States Air Force under Contract F49642-01-C-0003. Further information may be obtained from the Strategic Planning Division, Directorate of Plans, Hq USAF.

Library of Congress Cataloging-in-Publication Data

Future roles of U.S. nuclear forces : implications for U.S. strategy / Glenn Buchan ... [et al.].
p. cm.
Includes bibliographical references.
MR-1231-AF
ISBN 0-8330-2917-7
1. Strategic forces—United States. 2. United States—Military policy. 3. Nuclear weapons—United States. I. Buchan, Glenn C.

UA23 .F883 2000
355.02'17'0973—dc21

00-045817

RAND is a nonprofit institution that helps improve policy and decisionmaking through research and analysis. RAND® is a registered trademark *The views expressed in this report are those of the authors and do not reflect the official policy or position of the Department of Defense or the United States Air Force.*

Cover design by Tanya Maiboroda

Published 2003 by RAND
1700 Main Street, P.O. Box 2138, Santa Monica, CA 90407-2138
1200 South Hayes Street, Arlington, VA 22202-5050
201 North Craig Street, Suite 202, Pittsburgh, PA 15213-1516
RAND URL: http://www.rand.org/
To order RAND documents or to obtain additional information, contact Distribution Services: Telephone: (310) 451-7002; Fax: (310) 451-6915; Email: order@rand.org

PREFACE

This study examines the possible roles of nuclear weapons in contemporary U.S. national security policy. Since the end of the Cold War, the United States has been reexamining its basic assumptions about foreign policy and various instruments of national security policy to define its future needs. Nowhere is such an examination more important than in the nuclear arena.

Research for this document was completed in the summer of 2000 and, therfore, predates the current administration's Nuclear Posture Review. A lengthy governmental clearance and public release review process has resulted in the 2003 publication date of this formal report.

A lot has happened since then. The Bush administration has completed its NPR, which is classified, although much of it has been leaked to the press. The United States has conducted a war against Iraq, which it rationalized primarily on the grounds that Iraq was believed to be developing weapons of mass destruction (i.e., chemical and biological weapons in the near term; nuclear weapons in the long term). The United States also faces a confrontation with North Korea, which claims to have already developed a few nuclear weapons and threatens to make more, and Iran, which U.S. intelligence believes has a covert nuclear weapons program. The Bush administration has also announced plans to develop a new generation of nuclear weapons, improved earth penetrators with small-yield warheads to destroy underground facilities more effectively. The Bush administration has signed a new arms reduction treaty with Russia (i.e., the Moscow Treaty). It has also withdrawn from the

Antiballistic Missile (ABM) Treaty and announced its intention to deploy a National Missile Defense (NMD) system to protect the United States from attacks by rogue states. This report does not consider any of these specific events, although it does cover all the relevant general topics. Updating the report would amount to doing a whole new study, so we chose to release the report in its original form. The general analysis is still relevant and should inform any future debate on future U.S. nuclear strategy.

Futher, discerning readers will note a few locations in the text where, for reasons of classification, the authors have been forced to sidestep the historical record, and we beg the reader's indulgence for these instances. While they produce some distortion in facts as presented, they do not affect the basic analysis contained here. On balance, we judged that broader interests were served by the wide distribution of a slightly imperfect unclassified document, rather than more limited distribution of a classified report that would be more accurate in a narrow, technical sense.

This work should be of interest to those involved in nuclear strategy, force planning, arms control, and operational planning. The work was conducted in Project AIR FORCE's Strategy and Doctrine Program, which was directed by Dr. Zalmay Khalilzad at the time we did the work. Subsequently, Dr. Ted Harshberger succeeded Dr. Khalilzad as director of the Strategy and Doctrine Program. The project leader was Dr. Glenn Buchan.

PROJECT AIR FORCE

Project AIR FORCE (PAF), a division of RAND, is the Air Force federally funded research and development center for studies and analyses. PAF provides the Air Force with independent analyses of policy alternatives affecting the development, employment, combat readiness, and support of current and future aerospace forces. Research is performed in four programs: Aerospace Force Development; Manpower, Personnel, and Training; Resource Management; and Strategy and Doctrine.

Additional information about PAF is available on our web site at http://www.rand.org/paf.

CONTENTS

FIGURES

TABLES

SUMMARY

The defining characteristic of nuclear weapons—their almost unlimited destructive power—makes them unmatched as terror weapons and potentially more effective than any other type of weapon in strictly military terms (i.e., destroying targets). Moreover, the ability to produce nuclear weapons with relatively large yields in very small packages can dramatically increase their potential military value. Accordingly, nuclear weapons offer a range of strategic and tactical advantages to those countries that possess them. They can be used as instruments to

- coerce enemies by threat or actual use
- deter enemies from a range of actions by threat of punishment
- offset an imbalance of conventional forces
- fight a large-scale war
- destroy specific critical installations
- enhance national prestige and win a "place at the table" in the international arena.

The United States has used its nuclear forces for most of those purposes. Even more significant, it has *not* used them in combat since Nagasaki. Most notably, of course, the United States used nuclear weapons to coerce the Japanese to surrender in World War II and later maintained a large nuclear arsenal to deter the Soviet Union from launching a nuclear attack on the United States or invading Western Europe with its numerically superior conventional forces.

The United States also tried, with mixed success, to extract additional political mileage from brandishing its nuclear forces in peripheral conflicts.

The distinctive nature of the Cold War shaped the evolution of U.S. nuclear strategy and force structure in important ways. The dominant threat to the United States was the Soviet Union, an ideological adversary and competing great power armed with nuclear weapons that posed a direct threat to the United States after the Soviets developed long-range missiles and armies that appeared capable of overwhelming the conventional forces of U.S. allies in Western Europe. Once the Soviet Union developed intercontinental ballistic missiles (ICBMs) and submarine-launched ballistic missiles (SLBMs) armed with nuclear warheads, there was no way to protect the United States from a Soviet nuclear attack. After the Soviets deployed their missiles on nuclear ballistic missile submarines (SSBNs) and in hardened silos, disarming them with a nuclear first strike would have been virtually impossible, although the United States never stopped trying to develop the requisite technical capabilities. As a result, the best way to prevent a Soviet nuclear attack on the United States appeared to be to deter such an attack by threatening retaliation with U.S. nuclear weapons.

Implementing that deterrence strategy shaped U.S. strategic forces and operating practices in critical ways that affect U.S. forces to this day:

- A mix of ICBMs, SLBMs, and bombers—the so-called "triad"—was chosen in the 1950s to provide a diverse enough force to complicate an attacker's problem in trying to destroy the entire force and to hedge against technical failures of various sorts.
- A set of tactical warning systems and an associated network of command and control systems and procedures was developed to detect and characterize an impending nuclear attack on the United States, identify the attacker, and provide senior U.S. policymakers with at least a few minutes to respond to an attack before the system broke down.
- U.S. strategic forces were maintained at very high levels of alert—bombers on strip alert, SSBNs at sea, and ICBMs ready to launch within a few minutes—to minimize the effect of a surprise attack.

- U.S. weapons were pretargeted and integrated into a single massive plan—the Single Integrated Operational Plan (SIOP)—with a few variants to make execution of a retaliatory strike as simple, quick, and efficient as possible.

For its success, this approach depended to some degree on historical and geographic accidents:

- The time and space that separated the principal antagonists
- The time to develop and perfect intercontinental nuclear forces on both sides
- The relatively unique nature of those forces
- The relative simplicity of the largely bipolar world.

Although these factors helped reduce the stress and fog of the U.S.-Russian nuclear confrontation, *it was still very dangerous.* Because of the stakes in the competition (e.g., national survival), both sides were willing to take substantial risks—accidental or unauthorized launches, mistakes, miscalculations—to reduce their vulnerability to surprise attacks. Because of the sheer destructiveness of nuclear weapons, any mistake could have had catastrophic consequences. Everyone recognized that fact from the beginning and tried to take steps to reduce the dangers, but the perceived need to deter a deliberate nuclear attack took precedence.

The end of the Cold War changed a lot, but not everything:

- The dissolution of the Soviet Union and the Warsaw Pact greatly diminished the chances of general nuclear war or a major war in Europe. U.S. and Russian relations, while not exactly cordial since the post–Cold War "honeymoon" ended, are much less confrontational than in the past.
- U.S. and Russian nuclear forces are much smaller and operate at lower levels of alert. Still, Russian strategic nuclear forces remain the only current threat to the national existence of the United States. In addition to the overt threat, Russian economic woes; the deterioration of some of its nuclear forces, command and control and warning systems, and nuclear infrastructure; and the general failure of Russian economic and political reforms pose

new kinds of problems for U.S. security (e.g., nuclear theft, proliferation, and unauthorized use) and exacerbate old ones (e.g., war by accident or mistake).

- U.S. strategic nuclear forces are structured basically the same way they have always been. (U.S. tactical nuclear forces have largely been eliminated.) U.S. operational procedures have in the main changed little since the Cold War days.
- Nuclear proliferation is probably a greater problem now than it was during the Cold War. The odds of nuclear use by someone somewhere have probably increased.
- There may be more nuclear players and different types of players with different concepts of nuclear strategy and means of delivering weapons. *That situation could make defending against or deterring nuclear use more difficult.*
- Faced with U.S. military and economic dominance, other nations and nonstate actors are likely to seek different ways to counter U.S. power (e.g., terrorism, covert use of nuclear or biological weapons).
- Political instability in established nuclear states is a cause of major concern. *An established nuclear power coming unglued and lashing out is the worst possible threat to U.S. security for the foreseeable future.*

The United States is currently facing this world with a set of nuclear forces that is only a somewhat reduced version of the force it has maintained for decades. Similarly, its overall strategy is virtually the same—the only real difference is an explicit nuclear threat against countries developing biological and chemical weapons.

We found that the United States has a much broader range of nuclear strategies and postures among which it could choose, including at least

- abolition of U.S. nuclear weapons
- aggressive reductions and "dealerting"
- "business as usual, only smaller"

- more aggressive nuclear posture
- nuclear emphasis.

"Mixing and matching" is also possible. For example, a much smaller nuclear force operated differently could also be used more aggressively if the situation demanded it.

Devising a U.S. nuclear strategy for the future requires a mix of analytical assessments and value judgments. Among our key observations are the following:

- Nuclear weapons still lend themselves best to deterrence by threats of punishment, although one can never be certain how effective such threats will be. Even small nuclear forces should be capable of providing this kind of deterrence.
- Nuclear counterforce strategies, which would not have been effective during the Cold War, might actually work now, especially against emerging nuclear powers.
- The United States can influence, but no longer control, the nuclear "rules of the game" as it once did. As a result, it needs a wider variety of policy instruments than nuclear deterrence to deal with the range of potential nuclear threats.
- The degree to which the United States might need nuclear weapons for actual war-fighting depends to a significant degree on the demonstrated effectiveness of other kinds of forces (e.g., advanced conventional weapons, defenses).
- For most foreseeable actual combat situations, advanced conventional weapons are probably sufficiently effective *if the United States buys enough of them and uses them properly*.
- Still, nuclear weapons trump all others, and if the stakes were high enough, and other options were inadequate, nuclear weapons could give the United States a decisive advantage.
- Counterforce attacks against nuclear weapons that could reach the United States are an obvious example. Otherwise, only a situation where the United States was forced to fight a world-class opponent at long range and could not apply enough mass of firepower with conventional weapons might warrant the use of

nuclear weapons. That would probably require a large number of small nuclear weapons delivered by bombers. The United States does not now have such weapons.

- Unlike the Cold War, future situations that might require U.S. nuclear use are unpredictable. Thus, a *prerequisite for any strategy of nuclear use other than "set piece" exchanges with Russia is a flexibility in planning and execution that is the antithesis of the SIOP.*

- A strategy of deterrence and selective nuclear use could be implemented with a "dealerted" force, assuming that force was designed properly. *Nothing about deterrence by threat of punishment requires prompt retaliation,* and in an uncertain world, a hasty response could be more dangerous than in the past. Two assumptions are critical to the case for a dealerted force:

  — *The risk of accidental nuclear war must be viewed as greater than the risk of a surprise attack.*

  — *The Russians would react to a dealerted U.S. force by reducing their reliance on launch-on-warning and preemption.*

- The effect of U.S. nuclear strategy and force structure decisions on the likelihood of further nuclear proliferation is ambiguous and difficult to predict.

- Even if the United States wants to remain a major nuclear power, "withering away" of its nuclear capability over time may be inevitable. That would certainly be the most likely effect of continuing its current nuclear policies.

In sum, nuclear weapons remain the final guarantor of U.S. security. The United States has considerable flexibility in choosing an overall nuclear strategy for the future and in implementing that strategy. Among the range of options, a contemporary nuclear strategy that retains the traditional threat of nuclear retaliation in hopes of deterring serious threats to U.S. national existence coupled with the operational flexibility to actually use a modest number of nuclear weapons if the need is overwhelming and other options are inadequate may offer a balance of benefits and risks for as long as the United States chooses to retain nuclear forces. *Both the forces and the operational practices appropriate for enforcing such a strategy are likely to look very different from the current U.S. approach. Nothing about deterrence by threat of punishment requires prompt retaliation,* and in an uncertain world, a hasty response could be more dangerous than in the past.

ACKNOWLEDGMENTS

We particularly want to thank our colleagues Fritz Ermarth and Robert Nurick for their insightful comments on an early draft.

Several of our other colleagues were very helpful as well, and we appreciate their efforts. Jane Siegel typed the original manuscript. Sandra Petitjean and Mary Wrazen created many of the graphics. Alaida Rodriguez made further revisions to the text.

Emily Rogers did the painstaking work of making corrections to the final version of the text and completing the document.

ACRONYMS

ABM	Anti-Ballistic Missile
ADM	Atomic Demolition Munition
ASW	Anti-Submarine Warfare
BAT	Brilliant Anti-Tank
C^2	Command and Control
CBW	Chemical and Biological Weapon
CEP	Circular Error Probable
CONUS	Continental United States
CTBT	Comprehensive Text Ban Treaty
DGZ	Desired Ground Zero
DTRA	Defense Threat Reduction Agency
EMP	Electromagnetic Pulse
GBU	Guided Bomb Unit
GPS	Global Positioning System
HOB	Height of Burst
ICBM	Intercontinental Ballistic Missile
IR	Infrared
JSOW	Joint Standoff Weapon
km	Kilometer
kT	Kiloton
LEO	Low Earth Orbit

m	Meter
NATO	North Atlantic Treaty Organization
NPR	Nuclear Posture Review
NPT	Nuclear Non-Proliferation Treaty
PGW	Precision-Guided Weapon
PVH	Physical Vulnerability Handbook for Nuclear Weapons
rad	Radiation Absorbed Dose
RV	Reentry Vehicle
SFW	Sensor-Fuzed Weapon
SIOP	Single Integrated Operational Plan
SLBM	Sea-Launched Ballistic Missile
SRAM	Short-Range Attack Missile
SSBN	Nuclear Ballistic Missile Submarine
START	Strategic Arms Reduction Treaty
THAAD	Theater High-Altitude Area Defense
U.S. STRATCOM	United States Strategic Command
WCMD	Wind Corrected Munition Dispenser
WMD	Weapons of Mass Destruction
WR	Weapon radius

Chapter One

INTRODUCTION

Nuclear weapons are the ultimate guarantors of a nation's security. At least, that is what countries that possess them—or would like to possess them—believe. During the Cold War, a nuclear confrontation between the Soviet Union and the United States was the central reality in world politics. With the end of the Cold War and the dissolution of the Soviet Union, the world continues to evolve toward a more complex international order, less dangerous in some ways, perhaps more dangerous in others. During the Cold War, the most important threat to U.S. security, indeed to its very existence, was the possibility of a Soviet nuclear attack. Deterring such an attack was the central element of U.S. national security policy, and U.S. strategic nuclear forces were the primary instruments of that policy. Thus, nuclear forces were the centerpiece of U.S. national security strategy.

With the end of the Cold War, the perceived threat of a Russian nuclear attack—already considered to be low—diminished dramatically. Since then, both U.S. and Russian nuclear forces have been reduced substantially in size and readiness and have clearly moved to the "back burner" in discussions of critical national security issues and battles for funds and attention. There is a widespread view that nuclear issues no longer matter much for the United States. At the very least, there does not appear to be a clearly articulated view of why the United States still needs nuclear forces, what those forces need to be able to do, and what criteria an effective U.S. nuclear force needs to meet. In the meantime, U.S. nuclear policy and strategic force structure remain relatively unchanged, a combination of momentum and (relatively) benign neglect.

Such a policy is not sustainable indefinitely. If for no other reason, a series of decisions will be required to maintain, reduce, expand, modify, or even scrap various parts of the U.S. nuclear force. Political decisions will have to be made about formal arms control-related issues. Meanwhile, proposals to change U.S. nuclear policy are already on the table from people whose opinions matter. Proposals cover the spectrum from outright abolition of nuclear weapons to drastic cuts in force levels and radical modification of operating procedures to much more aggressive weapons development programs and operational concepts. The stasis cannot continue unabated. Sooner or later, the United States will require a new nuclear policy to provide a rational basis for future decisions on force structure and operational practice.

This study examines contemporary roles for U.S. nuclear forces and analyzes a number of alternative future U.S. nuclear strategies. Drawing on classical writings on nuclear strategy and the U.S. experience during the Cold War as well as subsequent work—both by RAND and others—on the changes in the wake of the Cold War, we show that the United States has a wide range of choices in crafting a contemporary nuclear strategy. Nuclear weapons will still retain their primary function of deterring by threat of punishment, although who is to be deterred from doing what to whom is even more problematic than in the past. Beyond that, the United States has considerable choice in how aggressive it wants to be in actually using nuclear weapons. We found that even a relatively small force operated much more flexibly than in the past could be both a deterrent and a war-fighting force if the stakes were high enough and all else failed.

Chapter Two reviews the basics of nuclear weapons—what they do, what their shortcomings are, and how U.S. nuclear strategy and forces have evolved over the 40 years of the Cold War. Chapter Three examines potential contemporary roles for U.S. nuclear forces and identifies some of the key issues that need to be resolved. In Chapter Four, we make some quantitative effectiveness comparisons between current nuclear and modern conventional weapons for selected applications to see in what situations, if any, nuclear weapons have an overwhelming advantage. Chapter Five looks at future U.S. strategic choices and addresses a number of specific issues that affect those choices. Chapter Six summarizes our conclusions.

Chapter Two

NUCLEAR WEAPONS AND U.S. SECURITY—BACK TO BASICS

In addressing the role nuclear weapons might play in contemporary U.S. national security policy, the first step is a "back to basics" review of nuclear weapons—what they do, what makes them unique, and how they have served U.S. security interests in the past.

WHAT NUCLEAR WEAPONS DO

The most fundamental characteristic of nuclear weapons is their almost unlimited destructive power. That destructiveness manifests itself in two ways. First is the potentially apocalyptic effects of a large-scale war fought with nuclear weapons. That, obviously, has been the driving force behind movements to reduce or eliminate nuclear weapons since the dawn of the nuclear age. Second is the enormous destructive power that can be put into a small package, which can then be delivered by any one of a number of means. A single nuclear detonation can destroy virtually any individual target or lay waste to large areas (e.g., destroy a city). That characteristic changed the nature of war dramatically. It appeared to make defense, in the traditional sense, virtually impossible because of the damage that even a single nuclear weapon that leaked through defenses could cause. Also, when coupled with long-range delivery systems (particularly long-range bombers and ballistic missiles), nuclear weapons allowed those possessing them to destroy an enemy's homeland without necessarily having to defeat its military forces first. Thus, nuclear weapons, if used effectively, could prevent an enemy's military from achieving the most fundamental objective of any military establishment: protecting its homeland. That changed the traditional concepts of war.

Even in strictly military terms, nuclear weapons are simply more effective than other weapons in destroying targets. Table 2.1 shows some classes of targets against which nuclear weapons are particularly effective. As experience with the weapons grew, so did the range of potential applications. Some took advantage of special effects of nuclear weapons other than just heat and blast.

Electromagnetic pulse (EMP) and radar and communications blackout are examples.

These characteristics of nuclear weapons offered attractive strategic advantages to those who owned them:

- Coercion of enemies by threat or use of nuclear weapons (e.g., the U.S. nuclear attacks on Japan to coerce Japan to surrender unconditionally and end World War II).
- Deterrence of a range of actions by threat of nuclear use.
- A means of offsetting an imbalance of conventional forces (e.g., the U.S. rationale for its nuclear posture in Europe; the original motivation for the Swedish nuclear weapons program, which never came to fruition).

Table 2.1

Targets for Which Nuclear Weapons Are Particularly Suitable

- Massed formations of troops, particularly armor
- Large military complexes (e.g., airfields, ports)
- Hardened military installations (e.g., missile silos, underground command centers)
- Inherently hard natural or man-made structures (e.g., concrete bridges or dams, cave or tunnel entrances)
- Large warships
- Arriving ballistic missile warheads
- Satellite constellations
- Some kinds of communications and electronic systems
- Industrial capacity and cities

- The most effective means for fighting any large-scale war.
- Prestige and a "place at the table" (e.g., the permanent members of the United Nations Security Council are the original five members of the "nuclear club").

The countries that have acquired nuclear weapons, or considered doing so, have emphasized different rationales and tailored the concepts to their own particular needs. The United States, for example, has taken advantage of all of these characteristics of nuclear weapons over the years in crafting its national security strategy.

Of all the types of so-called "weapons of mass destruction" (WMD), nuclear weapons are clearly the best to have, especially for countries with nuclear establishments already in place. As terror weapons, they are unmatched. As military weapons, they are more effective and more difficult to protect against than chemical, biological, or advanced conventional weapons.

RISKS AND DISADVANTAGES OF NUCLEAR WEAPONS

On the other hand, nuclear weapons have significant disadvantages and inherent risks as well, stemming mainly from the same characteristics responsible for their unique advantages. Primary risks include:

- excessive damage
- incidents, accidents, mistakes, and miscalculations
- unauthorized use
- theft
- operational difficulties
- "pariah" status
- increased proliferation
- environmental hazards and infrastructure problems.

The destructiveness of nuclear weapons has been a major concern since the beginning of the nuclear age. The fundamental concern has been that the damage from actual use of nuclear weapons would be out of proportion to any legitimate political or military ends. The danger has always been perceived as particularly acute in conflicts involving major nuclear powers owing to the sheer scale of the potential effects (e.g., large-scale fallout, climatic effects) of an unlimited nuclear exchange should one ever occur. Thus, escalation risk has been a major issue in superpower confrontations. However, even limited nuclear exchanges or nuclear use could fail the proportionality test inherent in the notion of "just wars" and reinforce the long-standing moral argument against nuclear weapons. That argument is likely to become an increased concern for the United States if the North Atlantic Treaty Organization's (NATO's) experience in Kosovo is any indication. Even precision conventional bombing of Yugoslavia did enough damage to raise arguments about the morality and effectiveness of coercive strategic bombing, rendering nuclear use almost out of the question in any but the most extreme circumstances.

The array of problems associated with risks of accidental use of nuclear weapons, incidents, false alarms, mistakes, miscalculations, and unauthorized use of nuclear weapons have long been the subject of discussion and scrutiny. The unraveling of the Russian nuclear establishment has exacerbated concern about some of these problems, including the danger of theft of nuclear weapons or nuclear material. Disagreements about the severity of these problems are at the heart of much of the current debate about future U.S. nuclear posture.

Aside from their effect on civilians, collateral effects of nuclear weapons can complicate military operations and cause a variety of headaches for field commanders. In addition to the obvious problems of operating in a radiation environment, there are more subtle difficulties as well. For example, nuclear detonations can black out some radars and communications systems, affecting all sorts of operations. For example, one of the problems with equipping antiballistic missile (ABM) interceptors with nuclear warheads is the concern over self-blackout of the tracking radars that can result, which could make it easier for subsequent attackers to penetrate the defense. As a consequence, nonnuclear ABM systems have always been attractive in principle on straightforward military grounds if they could be made to work. On a more mundane level, the problems associated with special handling of nuclear weapons, the need to obtain release authority to use them, and the competition for scarce support resources might convince military commanders that nuclear weapons are more trouble than they are worth unless the need is truly compelling. An interesting aspect of the current policy debate about nuclear weapons is the number of senior former military officers who have become disenchanted with nuclear weapons and are actively seeking ways to eliminate or drastically reduce them. *While much of the basis of their concern is clearly moral and political, there is also a strong operational flavor—e.g., option X is not practical or militarily sensible—that makes their opposition to nuclear weapons particularly compelling.* Field commanders have voiced these complaints for decades. These complaints not only resonate at the operational level, but they also have fundamental implications for grand strategy.

Finally, the flip side of the argument that nuclear powers acquire a heightened status is that they might also be regarded as pariahs. This has always been a delicate balancing act for the established nuclear powers, which are obliged by the Nuclear Non-Proliferation Treaty (NPT) to move toward nuclear disarmament. New nuclear powers—the few that there have been—have not obviously improved their stature in the international community by demonstrating their nuclear capability. India and Pakistan certainly have not. Indeed, both have suffered economic and political sanctions as a result of their nuclear tests, and both are arguably less secure than they were before. By contrast, Israel has always found it more effective to be an "undeclared" nuclear power, deriving deterrent value from the universal perception of its nuclear capabilities without having to pay the political price that becoming an overt nuclear power would entail.

THE HISTORICAL LEGACY

Over the 40 years of the Cold War, the United States developed nuclear forces, operating procedures, and a strategic view of nuclear weapons that reflected the needs and possibilities of the times. Those experiences will invariably shape—for better or worse—U.S. perspectives on contemporary nuclear strategy. Reviewing where we have been is a prerequisite to deciding where we want to go and how best to get there, particularly because U.S. nuclear policy has been in stasis since the end of the Cold War. Indeed, that policy has largely been on autopilot since the Cold War days. That is not entirely bad—some of the aspects of past U.S. nuclear policy probably are transcendent. However, as we will argue later, that momentum cannot go on indefinitely. Absent some movement, U.S. nuclear policy will become one of "withering away by default"—the gradual deterioration of U.S. nuclear capability because no one is minding the store. *"Withering away" by design might be an acceptable policy. Withering away by default could be dangerous.*

Briefly, the Cold War world was a much simpler place:

- It was dominated by two ideologically opposed, nuclear-armed, major military and political powers, the United States and the Soviet Union.
- Being on opposite sides of the world, the United States and the Soviet Union were separated by time and space. That time and space would help reduce the friction and eventually provide enough warning time to allow a stable strategic nuclear balance to develop between them, in spite of direct confrontations in places such as Berlin and indirect confrontations in Korea and elsewhere.
- The maturing of long-range delivery systems and nuclear warhead technology also took some time, which provided both sides a cushion to learn how to coexist.
- The world itself was largely polarized into competing camps, although the degree of bipolarity can be overstated. The bipolar alliances both exaggerated the importance of minor conflicts and increased the risk of U.S.-Soviet confrontation, and simultaneously placed some constraints on the behavior of superpower allies.
- In spite of constraining factors, the first decade and a half of the U.S.-Soviet nuclear confrontation were very dangerous, perhaps even more so in retrospect.[1]

U.S. nuclear forces and strategy evolved over the course of the Cold War. Key elements of that evolution included the following:

---

[1]Craig (1998), Allison and Zelikow (1999), and Trachtenberg (1991) expand on this topic.

- Early theoretical work in the late 1940s, based on first principles rather than empirical evidence, suggesting that the main function of atomic bombs might be *deterrence* rather than actual use. This work had little, if any, practical impact on policy at the time (although, ironically, President Truman seemed to understand the point clearly), but it did put down an intellectual marker.
- The dramatic U.S. nuclear weapons buildup that began in the 1950s focused heavily on a massive strategic nuclear bombing campaign aimed at destroying Soviet military capability. This was a direct application of the strategic bombing doctrine honed by the United States and others in World War II. The operational practices at the time were most appropriate for executing a preemptive attack (i.e., striking first).
- There was a simultaneous large-scale buildup of a wide variety of U.S. tactical nuclear weapons. These included everything from atomic demolition munitions (ADMs) intended mainly to blow up bridges, air defense missiles, torpedoes, and depth charges to the more familiar bombs, artillery shells, and shorter-range missiles.
- Based largely on empirical evidence about the way U.S. bomber forces operated, Albert Wohlstetter described the need to be able to strike second to deter enemies from launching nuclear attacks and the practical difficulties (e.g., survivability of forces, adequate command and control) in doing so (Wohlstetter, 1959).
- At about the same time, Herman Kahn presented a more fine-grained view of deterrence, describing levels of actions that the United States might seek to deter.[2]
- Meanwhile, weapons development decisions during the Eisenhower administration defined the strategic nuclear force structure—ICBMs, sea-launched ballistic missiles (SLBMs), and long-range bombers—that the United States maintains to this day.

---

[2]Kahn (1969, p. 126) defined Type I deterrence as deterrence of a direct nuclear attack on the United States; Type II deterrence as deterrence of very provocative acts other than a direct attack on the United States itself; and Type III deterrence as deterrence of lesser provocations.

- In the early 1960s, U.S. Secretary of Defense Robert McNamara, after flirtations with strategic doctrines emphasizing counterforce and city-avoidance, committed the United States for the first time to a doctrine of deterrence rather than war-fighting, which implied upper limits on the size of strategic nuclear forces that were needed. This sea change codified in policy for the first time that nuclear weapons were not actually intended to be used. Counterforce was viewed as counterproductive according to this logic; so was strategic defense.
- The emphasis on maintaining the capability to respond to a nuclear attack rather than initiating one became the focus of most attention in the U.S. defense community for the last three decades of the Cold War. Forces and command and control systems had to survive long enough to launch a successful second strike, and warning systems had to be able to warn of the impending attack and identify the attacker in time to allow a retaliatory response.
- Strategic targeting over this period was relatively unaffected by any of these doctrinal debates. The United States always targeted a comprehensive set of Soviet and other military nuclear and conventional forces. There were nuanced changes, but continuity was the rule.
- The operational change that *did* matter was the creation of the Strategic Integrated Operational Plan (SIOP) in 1960. The SIOP was intended to bring some order to the targeting process and integrate the burgeoning nuclear forces of the different services into a single, efficient plan.[3] The SIOP was designed to solve a small number of "set piece" problems—defeating or responding to a Soviet nuclear attack on the United States. The fear at the time was that the SIOP might have to be executed under the extreme pressure of a massive nuclear attack. Thus, it had to be simple, relatively rigid, and preplanned in exquisite detail. The basic planning assumptions have changed relatively little over the years, although emphasis has shifted occasionally. The SIOP is still with us today.

---

[3]The first SIOP was also a victory for the Air Force in its internecine battles with the Navy over who controlled strategic nuclear targeting.

- Since the early 1960s, technology has changed, weapons systems have improved, and the whole process of operating strategic nuclear systems has become more refined. However, the overall patterns of strategic force development and planning have remained largely unchanged.
- Beginning in the late 1960s, formal strategic arms control became an institutionalized part of the strategic nuclear force calculus and policy process. That continues to the present day, although its future is uncertain.
- Perhaps most important, no nuclear weapon has been used in anger since Nagasaki, in spite of some near misses (e.g., the Cuban missile crisis).

This rehash of Cold War history defines the starting point on the game board that U.S. planners currently have to work with. Key elements of U.S. nuclear policy have been remarkably resilient over the years. Most important are the tradition of non-use of nuclear weapons, the strategic "triad," the SIOP, the emphasis on striking second (although striking first has never been precluded), the focus on deterrence by threat of punishment, the role of formal arms control in the strategic planning process, and the virtual elimination of strategic defenses. *The issue for contemporary U.S. nuclear planners is whether the momentum of past policies should be maintained or whether some or all of the key elements should be modified, replaced, or discarded.*

It is also worth reflecting on what U.S. nuclear policies accomplished during the Cold War. Unfortunately, relating cause and effect is virtually impossible. Still, we can examine what actually *happened* and, in some cases, draw plausible inferences.

- There was no general war between the United States and the Soviet Union. Indeed, there was no nuclear use (i.e., detonations in anger) of any kind.
- There was no global-scale war of any kind, in spite of the ideological conflict and the competition among great powers. That is a considerable improvement over the first half of the 20th century.
- There was no war in Europe despite the NATO-Warsaw Pact friction.

- There were peripheral wars involving the superpowers (e.g., Vietnam) in which nuclear capability conferred no advantage at all.
- There were other regional conflicts involving either the superpowers directly or their traditional allies where the potential influence of U.S. nuclear weapons was more ambiguous in influencing the outcome (e.g., Korea, the Middle East).
- Only a modest amount of nuclear proliferation occurred.
- Conflicts involving superpower allies and client states did occur, but none got completely out of control.

In sum, the United States achieved all of its major political objectives during the Cold War without sacrificing any vital interests. U.S. nuclear capability almost certainly played some role in that success. On the other hand, there were clearly limits to how much political utility nuclear weapons had in situations where the stakes were lower and the direct relevance of nuclear weapons was less clear. If Herman Kahn were keeping score, he probably would have concluded that Type I deterrence worked, Type III deterrence failed, and Type II deterrence was ambiguous, probably working to some degree in some cases.

Chapter Three

CONTEMPORARY ROLES FOR U.S. NUCLEAR WEAPONS

The Cold War has been over for ten years now, and the world has moved on. A lot has been written about what that means for U.S. nuclear force posture. Two interesting trends have emerged so far. First, there is widespread agreement among quite disparate parts of the defense community on some important issues. In particular, nearly everyone agrees that U.S. nuclear forces can be reduced drastically compared to Cold War levels. Second, even that degree of consensus masks remaining deep philosophical differences on the most basic concepts of nuclear strategy, the nature of the risks, and practical steps that the United States should take with respect to its nuclear forces. At that point, the consensus breaks down completely. Several proposals for different courses of action from serious people whose views matter are currently on the table in the public arena, and they demonstrate where nuclear views diverge. Thus, the United States has practical choices to make about nuclear force issues, and the public dialogue that has begun provides U.S. policymakers with an opportunity to review U.S. nuclear policy.

THE NEW SECURITY ENVIRONMENT

Description of the global security environment in the wake of the Cold War has become a standard litany. Still, it is fundamental to understanding the problems that U.S. nuclear forces might be called upon to help solve.

Future Nuclear Threats

The most important and dramatic change remains the dissolution of the Soviet Union and the Warsaw Pact and the ratcheting down of the likelihood of a nuclear confrontation between the nuclear superpowers and their allies. Russia is much less of a military and ideological adversary than it used to be, although the immediate post–Cold War euphoria about warmer U.S.-Russian relations has abated to a significant degree. Russia's conventional military forces have eroded substantially and pose little threat to others for the foreseeable future. Russia has suggested publicly that it will compensate by relying more heavily on nuclear weapons (presumably tactical nuclear weapons in particular) to protect its borders. Russia's nuclear forces have decreased dramatically as well, partly as a result of agreements with the United States and unilateral actions in the aftermath of the Cold War, and partly because of the state of the Russian economy. Even if Russia had not ratified the Strategic Arms Reduction Treaty (START) II, its strategic nuclear forces will drop below the levels allowed by the treaty because Russia will not be able to afford to maintain a force of that size. Moreover, alert levels have dropped as well. Thus, the sheer magnitude of the potential Russian nuclear threat to the United States has already decreased dramatically and will continue to decrease for at least a number of years. More important, even given the somewhat soured relations between Russia and the United States as the 20th century drew to a close, there is no quarrel between them that appears sufficient to provoke a nuclear war in the foreseeable future. As a result, the dominant threat to U.S. society for the last half-century—and the primary raison d'être for its nuclear arsenal—has certainly diminished in importance, although it still exists physically.

That said, even a massively reduced Russian strategic nuclear arsenal remains the only military threat capable of utterly destroying U.S. society. As a result, the United States still needs to be concerned about how to reduce the potential risks that Russian forces pose. There are several aspects to the problem. First, if the Russians still feel the need to be able to respond to a surprise attack on their nuclear forces (presumably from the United States), they may feel obliged to take extraordinary measures to protect their ability to respond. Continuing their past policies of launching vulnerable ballistic missiles on tactical warning of an attack would be bad enough, particularly considering the deteriorating state of Russian missile warning systems. Even worse, the Russians might take the extreme step of opting for a preemptive attack if they even *thought* they were about to be attacked first.[1] Obviously, such a posture would be very dangerous for both the United States and Russia because of the inherent risks of accidents or mistakes.

Such accidental nuclear war has been a long-standing nightmare of nuclear strategists. *The relative likelihood of accidental nuclear war and the importance of reducing its risks are among the major differences separating opposing schools of thought on contemporary nuclear strategy. So is the degree to which alterations in U.S. nuclear force posture and operational practice can reduce these risks by influencing Russian perceptions and behavior.*

Second is the ability of the Russians to retain centralized control of their strategic nuclear weapons, particularly if political chaos erupts. Russia has traditionally placed more emphasis than the United States on centralized control of strategic nuclear forces, so it may be better positioned to handle such problems if they should arise than other nuclear powers. Still, this is the kind of issue that could have come up when the United States and Russia were negotiating with Ukraine, Kazakhstan, and Belarus to give up their nuclear weapons. If an ICBM had been launched from one of those nuclear-capable former Soviet republics, for example, one question would have been, "Whose finger was on the button, and what did he hope to gain by this act?" Thus, attributing blame for an attack might have been more difficult than in the Cold War days when identifying the country of origin of a missile launch was generally considered sufficient. This makes retaliation more problematic, which could undermine deterrence that is based on threats of retaliation. Obviously, that kind of problem could arise again if Russia or any other nuclear power started to come "unglued."

Fortunately, this kind of bizarre scenario is probably unlikely. However, it is probably *less unlikely* in the new world than in the old. As Quinlivan put it:

[1]Russians raise this possibility occasionally in conversations with U.S. analysts. It is unclear whether they believe that Russia would actually adopt such a dangerous policy, or whether they view it merely as a bargaining chip to influence U.S. decisions.

> The first casualty of any nuclear weapon use should be the presumption that we know exactly what is going on.[2]

Finally, there is an even worse possibility—that of any established nuclear power coming unglued and lashing out, for whatever reason, with nuclear weapons. The fact that such an act would not be "rational" is precisely the point. A nuclear power with nothing left to lose might not be deterred by threats of retaliation. *This is the worst imaginable nuclear scenario.* It is much worse than the so-called "rogue nation" threat because the attack is likely to be larger and more competently executed than an attack by a newcomer with a limited nuclear arsenal.

The only other current nuclear power that is generally considered a potential threat to the United States itself is China. China and the United States have a complex relationship that has been complicated by recent events, including

- renewal of the perennial dispute over Taiwan, including among other things a ham-handed threat by a Chinese official to "nuke Los Angeles" if the United States interfered[3]
- allegations of U.S. companies illegally transferring missile technology to China
- allegations of Chinese espionage that provided China with detailed information on all U.S. nuclear weapon designs
- the U.S. bombing of the Chinese embassy in Belgrade during NATO's operations in Kosovo, and the Chinese insistence—apparently widely believed by the intelligentsia, all strategic logic to the contrary notwithstanding—that the attack was deliberate.

Whatever the truth of the stories about Chinese spying, China may have the most dynamic nuclear weapons program of any established nuclear power and is now capable of building modern ballistic missiles that can reach the United States and probably even defeat a rudimentary ballistic missile defense system. So far, Chinese nuclear

---

[2]Quinlivan and Buchan (1995), p. 12.

[3]This threat was widely reported in the media at the time (AFX News, 1996) and found its way into the *Congressional Record* (Ehrlich, 1996).

forces are relatively small. Future force sizes are, of course, uncertain, but U.S. threats to build a national ballistic missile defense system could lead the Chinese to expand their forces.

The Chinese nuclear threat to the United States could evolve into a smaller version of the former Soviet threat. The United States has experience dealing with that kind of situation. However, there could be significant differences. The problem is that the Chinese might not understand the "rules of the game" the way the United States does. That is one interpretation of both the crude Chinese nuclear threats over the Taiwan crisis and Chinese views of the U.S. bombing of the Chinese embassy in Yugoslavia. If the strategic views of China and the United States are really fundamentally different, a collision that no one wants could result from misunderstanding and miscalculation.

Another possibility that could become an issue if U.S. and Russian strategic levels drop substantially is the formation of coalitions among nuclear powers. The Soviet Union always used to raise that issue with the United States in earlier arms control negotiations, claiming that it should be allowed some extra margin to account for the nuclear arsenals of U.S. allies Britain and France as well as China. However, such arguments hardly mattered when the United States and the Soviet Union each had thousands of nuclear weapons and the other players had only a handful. If U.S. and Russian arsenals were cut drastically to, say, several hundred warheads each and the other nuclear powers maintained their arsenals at about their current sizes, coalitions of nuclear powers could start to matter numerically. How best to achieve a stable nuclear balance in that kind of world is a question that is receiving analytical attention at the moment and could become a real policy issue if the United States were to actually consider very deep reductions in its nuclear forces.

Beyond the usual players in the nuclear arena, others could emerge. Table 3.1 shows several classes of potential nuclear actors.[4] The striking feature is the rich variety of possibilities. Some might be able to threaten the United States directly; others may not, but could be serious threats in regional conflicts. Some are likely to know what

---

[4]Table 3.1 is adapted from Glenn Buchan, *Nuclear Weapons and the Future of Air Power,* RAND (forthcoming).

Table 3.1

A Scorecard for Evaluating Nuclear-Armed Opponents

	Characterizing the Nuclear Threat[a]				
Class of Opponent	Can Weapons Reach the U.S.?	Force Size	Force and C3 Vulnerability	Operational Competence	Societal/ Organizational Vulnerabilities
Major nuclear power	Yes	Large	Low	High	Usual
"Degraded" major nuclear power	Yes	Large	Low, but degrading	High, but degrading	Usual
Minor nuclear power	Yes	Small to moderate	Varies	Varies	Usual
Regional nuclear power	No	Small to moderate	Varies	Varies	Usual
"Symbolic" nuclear power	Probably not	Very small	High	None	Usual
"Virtual" nuclear power	No (?)	None-in-being to small	Varies	None	Usual
Nonnation state • with territory or other physical assets	?	Very small	?	Low by usual standards, but . . .	Variation of the usual
Non-nation state • with mainly "virtual" assets	?	Very small	?	?	Leaders, bank accounts, linkages
Terrorists/ nihilists	?	Very small	?	?	Leaders, members, limited assets

[a]The nature of the nuclear threats and the likely effectiveness of alternative strategies for dealing with them will vary dramatically among the different classes of opponents.

they are doing—for example, how to control, operate, and use the weapons effectively. Others are likely to be only marginally competent, even clueless.

Conventional wisdom suggests that regional nuclear powers may become more of a concern than they have been in the past. The old Cold War alliances provided the major nuclear powers with some measure of control over regional allies at the risk of drawing the major powers into conflict over peripheral quarrels (e.g., the 1973 Middle East war). With the end of the Cold War, the Russian and U.S. nuclear "umbrellas" tended to shrink, perhaps persuading some regional powers that they might be better off developing nuclear weapons of their own. On the other hand, even that sword cuts both ways. The United States was apparently willing to turn a blind eye toward Pakistan's efforts to circumvent technology transfer restrictions intended to inhibit nuclear proliferation because it needed Pakistan's assistance in aiding Afghan rebels fighting a guerrilla war against the Soviet Union.[5] Absent the Cold War, it is unlikely that the United States would have cared about the war in Afghanistan, certainly not enough to permit a nominal ally to acquire nuclear weapons.

The need to balance competing policy interests—a U.S. desire to prevent or limit nuclear proliferation in general versus the need to maintain the crucial support of a nominal ally in a Cold War battle—has resulted in a certain U.S. ambivalence toward nuclear proliferation.

That ambivalence is certainly consistent with the history of U.S. nonproliferation policy in general. As Nye observed, of the four post-NPT proliferators—Israel, India, Pakistan, and South Africa—only South Africa, a pariah state for other reasons, was subjected to severe sanctions for developing nuclear weapons,[6] and even those were ex post facto punitive rather than preventive measures. Nonproliferation is only one objective of American foreign policy, and in the

[5]Coincidentally, former senior U.S. diplomat and arms control negotiator Gerard Smith used exactly the same expression—"turned a blind eye"—in 1989 to describe the U.S. attitude toward both Pakistan's and Israel's development of nuclear weapons. (Cited in Federation of American Scientists, [1998].)

[6]Blackwill and Carnesale (eds.), p. 79.

case of Pakistan, the United States gave higher priority to defeating the Soviet Union in Afghanistan than to nonproliferation.[7] The United States dealt with Israel in a similar fashion during critical stages of its nuclear weapons program.[8] In the specific case of Pakistan, the United States only appeared to be serious about applying sanctions in the middle-to-late 1970s before the Soviet invasion of Afghanistan in 1979 and after 1991, when Pakistan's violation of U.S.-imposed constraints on its nuclear program (i.e., enriching uranium beyond the 5 percent limit asserted by the United States to be acceptable) became too blatant to ignore and—not coincidentally—the Soviet Union had withdrawn from Afghanistan and the Cold War had essentially ended.[9] In the interim, Congress and a series of administrations sparred over sanctions to inhibit or punish Pakistan's development of nuclear weapons.[10] Indeed, the very fact that Congress felt the need in 1995 to pass the Pressler Amendment, which made military aid to Pakistan contingent on the certification by the president that Pakistan did not "possess a nuclear explosive device," makes a strong a priori case that the United States was ambivalent at best about restraining Pakistan's nuclear program.[11] Others went further and accused the United States of actual connivance in assisting Pakistan in developing nuclear weapons.[12] These kinds of competing pressures on national governments are precisely the kinds of factors that make controlling nuclear proliferation difficult, particularly when the potential proliferators are bound and determined to acquire nuclear weapons for what they consider to be vital national security reasons.

The emergence of nonnation-state actors has long been considered a possibility, albeit a remote one. The end of the Cold War might make that more likely, although even now such groups might well consider acquiring nuclear weapons to be more trouble than it is worth. If such groups were to obtain nuclear weapons, the threat of nuclear

---

[7]Ibid.; Perkovich (p. 265); Spector (p. 129).

[8]Cohen (especially, pp. 193–196).

[9]Spector, pp. 121–125; Blackwill and Carnesale, p. 111.

[10]Federation of American Scientists (FAS) (1998); Spector, pp. 126–148.

[11]FAS (1998); Blackwill and Carnesale, p. 111.

[12]Chopra and Gupta, p. 5.

retaliation might be ineffective or inappropriate to deter them from nuclear use. Unlike nation states, such groups might have nothing to threaten directly with nuclear weapons (e.g., no land to target). On the other hand, they might be vulnerable to other kinds of threats (e.g., confiscation of financial assets, assassination of the group's leaders). The general point is that if such threats were to emerge, traditional concepts of nuclear deterrence and defense might have to be broadened.

Some nuclear powers may use unconventional delivery means (e.g., nuclear devices in trucks or holds of ships), either because they have no alternative or because unorthodox approaches are their method of choice.[13] That could make both defending and responding much more complicated. There could be no tactical warning, and identifying the attacker could be more difficult, at least for isolated nuclear attacks. For example, North Korea's thinly veiled threat to respond to any U.S. military action against it by smuggling nuclear bombs into the United States and detonating them covertly even after having "lost" a war might well give U.S. planners pause. Indeed, Builder has argued that nuclear weapons may become weapons of the "weak," rather than the military "crown jewels" reserved for rich, scientifically advanced countries.[14] Similarly, by combining selected technologies associated with the so-called "revolution in military affairs" (e.g., Global Positioning System [GPS] navigation, compact guidance systems), readily available vehicle technology (e.g., small

---

[13]The recent claims by a former high-ranking Russian military officer that the Soviets included nuclear suitcase bombs in their repertoire of strategic weapons to attack key targets in the United States are intriguing in this regard. This story received considerable media attention in a somewhat different context when retired General Alexander I. Lebed told a visiting U.S. congressional delegation and Western reporters that 100 of the suitcase bombs were missing and could have found their way into the hands of terrorists (Paddock, 1997). The strategic implications of the original deployment of the weapons received less attention but raises interesting questions. Apparently, nuclear weapons were to be smuggled into U.S. cities by agents and detonated on command in some (unspecified) fashion as part of an overall nuclear attack on the United States. While such ideas have always been around, western nuclear strategists have tended to dismiss such delivery techniques as overly risky and unreliable, particularly given that the major nuclear powers had other delivery means available (e.g., missiles, aircraft). (By contrast, the United States planned to use atomic demolition munitions (ADMs) very differently, and even then they were considered peripheral to other types of nuclear weapons.) If the story about Soviet nuclear "suitcase" bombs is true, it could cast this option in a whole new light.

[14]Builder (1991).

cruise missile airframes), and rudimentary nuclear warhead technology, emerging nuclear powers might be able to develop quite respectable nuclear weapons to use against their neighbors, "visiting" major powers, or even the U.S. homeland in less time than historical evidence would suggest. After the Gulf War, an Indian defense official observed that the main lesson for regional powers of the war with Iraq was that the only way to take on a superpower and win was to have nuclear weapons.[15] He could be right. Ironically, the world may have indeed turned to the point where regional powers, and even Russia, view nuclear weapons as their most effective potential counter to American conventional superiority. The question, then, is what role, if any, U.S. nuclear weapons could play in that equation.

There is also the possibility that emerging powers may acquire nuclear weapons with the idea of actually *using* them instead of merely brandishing them,[16] particularly in cases of historic quarrels between neighboring states where territorial disputes, ethnic and religious battles, and plain old-fashioned hatred are involved. They may not accept the "rules of the game" that have evolved in the competition among the established nuclear powers.

Even if regional powers want to establish stable balances as the Soviet Union and the United States did during the Cold War, geography and the state of technology could make that extremely difficult for most regional competitors. Distances—and, therefore, missile flight times—are too short to allow the sort of survivable postures for land-based nuclear systems that the United States and the Soviet Union were able to achieve, even if the countries were to invest in suitable tactical warning systems.[17] Even air attacks could probably achieve tactical surprise against most regional powers because most have relatively porous air defense systems. Since offensive missiles and aircraft suitable for preemptive nuclear strikes are widely available, countries threatened by nuclear-armed neighbors will not have

---

[15]Mohan (1999).

[16]See Quinlivan and Buchan (1995).

[17]India and Pakistan are an obvious case in point. Their recent round of nuclear tests reduced the security of both nations, and their border clashes continue. The military coup in Pakistan probably will not help either. The two countries will have to find a different approach than the one used by the United States and the Soviet Union if they want to reduce the risks of nuclear war.

much time to work out a defense, even if they have nuclear weapons of their own. New powers could opt for mobile missiles operated routinely out of garrison or, if geography permitted, submarine-based missiles. However, these types of forces are expensive and difficult to operate, particularly the first time out with nuclear weapons.

Regional nuclear arms races of the sort that India and Pakistan are now engaged in will almost certainly increase the risk that nuclear weapons will be used again in anger by someone somewhere.[18] Such an event, and the world's reaction to it, could change history. If nuclear weapons were used in a populated area with the inevitable massive damage, world revulsion might reinforce efforts to abolish or radically constrain nuclear weapons. On the other hand, if the weapons were used in a remote area with relatively little civilian damage and had a decisive military effect, the reaction might be very different. Nuclear use might be legitimized, at least to some degree, and nuclear weapons might again be viewed as "winning weapons."

The above is one of several reasons why the United States has a strong interest in minimizing further proliferation of nuclear weapons. Because nuclear use of any kind by anybody for any reason might legitimize subsequent use by others, nobody has so far found any provocation adequate to justify that risk. The United States is a particular beneficiary of this nuclear non-use policy in that it has the most to lose in a nuclear war.

Beyond the danger of nuclear use in general, there is the obvious problem that the United States might choose to involve itself in a regional quarrel in which one or more of the antagonists has nuclear weapons. That would, of course, increase the risks to U.S. forces in

[18]The only real chance that India and Pakistan, for example, have at this point to climb out of the hole which they have dug for themselves and establish some measure of stability in their nuclear relationship is to try to become the "virtual" nuclear powers that some analysts predicted they wanted to be in the first place. That would mean dispersing the components for the weapons and hiding them, so the weapons would be neither immediately threatening nor readily vulnerable to attack. Practical problems abound, but the idea is that the nuclear weapons would have deterrent value because they could be assembled and delivered in spite of an attacker's best efforts to prevent it.

the area considerably and make it more difficult to achieve the military or political objectives the United States has in the region.

What emerges is the need for a richer concept of deterrence and defense and a more diverse set of tools to use to counter nuclear threats in the future:

- Deterrence of nuclear attacks by threat of nuclear retaliation may continue to be the best option against the standard nuclear threats of the past.
- Other kinds of deterrent threats or defensive concepts may be needed to cope with other kinds of nuclear threats.
- Deterrence of any sort may be ineffective against some kinds of nuclear threats, so some other approach will be needed to deal with them.

Other Kinds of Threats

Fortunately, most threats to U.S. security or situations in which the United States gets involved militarily do not and are not likely to involve nuclear weapons. One of the rationales for the original U.S. nuclear buildup was to counter the overwhelming Soviet conventional advantage in Europe. With the end of the Cold War and the disintegration of Russia's conventional military forces, that particular problem has disappeared for the foreseeable future.

The NATO example is interesting on several levels. U.S. nuclear strategy regarding NATO has always been fraught with ambiguity, which was probably necessary, as a practical matter, and required considerable finesse to slide by the logical problems and inherent contradictions. Initially, the United States relied on its strategic nuclear superiority over the Soviet Union to offset NATO's conventional weakness. The United States and its war-weary European allies did not feel they could afford to match the Warsaw Pact's conventional forces (and they did not relish the prospect of another major conventional war in Europe in any case), so they depended on the threat of U.S. nuclear retaliation against the Russian homeland to deter a Soviet invasion of western Europe. Later, NATO supplemented that strategy by deploying several thousand tactical nuclear weapons in Europe to offset the Warsaw Pact's advantage in conventional weapons and to increase the credibility of NATO's threats to use nuclear weapons in response to an invasion of western Europe.

Of course, the credibility of the U.S. threat of nuclear retaliation against the Soviet Union in response to a Soviet invasion of western Europe was called into serious question when the Soviet Union developed strategic nuclear forces of its own. That problem was lost on no one at the time, and NATO policy for the last several decades of the Cold War was a high-wire act designed to make it as difficult as possible for the United States to "sit out" a war in Europe. The rather tortured and Byzantine logic necessitated by reconciling the conflicting interests of the United States and its European allies basically relied on creating an escalation process that would make general nuclear war very difficult to avoid following a Russian conventional attack on Europe, in spite of all the rational incentives the principal antagonists had to avoid such an outcome. As noted earlier, the strategy worked, for whatever reason. There was no war, but the process was very dangerous.

The irony is that NATO's declaratory nuclear policy remains unchanged in the wake of the Cold War. In principle, NATO still relies on U.S. strategic nuclear retaliation to deter attacks on any of its members. Extending the U.S. nuclear umbrella to include new NATO members—the first wave made up of a reunified Germany, Poland, Hungary, and Czechoslovakia—was a contentious issue at the time of the NATO expansion decision. NATO countries probably did not take the problem all that seriously because a Russian invasion seemed out of the question and the political environment at the time seemed so much more relaxed than during the Cold War days. However, Russian concerns about NATO nuclear policy were probably more than just rhetorical, particularly considering the possibility that future NATO expansion might include the Baltic states or other former Soviet republics. Subsequently, NATO's involvement in Kosovo probably confirmed long-held Russian fears that NATO was more than a "defensive" alliance. The deteriorating political relations between Russia and the United States and the rest of NATO suggest the danger of further friction in the future.

So far, there has been little reason to worry about resolving the residual ambiguities for NATO nuclear policy. The United States has withdrawn all but a handful of its tactical nuclear weapons.[19] Those remaining are all gravity bombs, presumably to be delivered by whatever dual-capable nuclear aircraft remain in Europe. What military problems they could actually solve remains manifestly unclear. To be of any practical use, the weapons require operators to train enough to maintain their nuclear proficiency and planners to understand nuclear weapons well enough to know how and when to use them. Moreover, since the aircraft in Europe are even more vulnerable than during the Cold War to either preemptive attack on their bases or modern enemy air defenses, and since they have limited range, the kinds of military situations in which they could actually be used effectively is severely limited. Conditions would have to be dire enough than conventional forces would be inadequate to solve the problem, but the enemy would have to be incapable of either attacking U.S. air bases or nuclear storage sites directly or shooting down current, nonstealthy U.S. fighters before they could drop their bombs. Also, of course, the targets have to be in range of U.S. fighters operating from European bases. Such a scenario may not be impossible, but it would certainly be bizarre. The purpose of the U.S. tactical nuclear weapons currently deployed in Europe is political, not military. They appear to be adequate, so far at least, to reassure the European members of NATO of the U.S. nuclear commitment to them without being as provocative and potentially dangerous as

[19]During the Cold War, the United States maintained over 7,000 tactical nuclear weapons in Europe, plus additional naval weapons of various sort on naval vessels that could be committed to Europe's defense. The weapons were quite diverse, including small nuclear weapons to be implanted by humans (Atomic Demolitions Munitions [ADMs]), large numbers of nuclear artillery shells of various sizes, nuclear gravity bombs for dual-capable aircraft, and several different classes of nuclear-armed missiles. Moreover, nuclear weapons were an integral part of all of NATO's war plans. NATO planning staffs included analysts with considerable nuclear expertise. Operators trained routinely for nuclear missions, and nuclear use was included in NATO exercises.

As a result of the Intermediate Nuclear Forces (INF) Treaty between the United States and the Soviet Union and unilateral actions by the United States (generally in response to Soviet and later Russian initiations as the Cold War ended and the Warsaw Pact collapsed), the United States removed all of its land-based nuclear missiles, all of the Army's nuclear weapons, and all non-strategic naval nuclear weapons from Europe. Only nuclear gravity bombs remained. Thus, the remaining nuclear force in Europe is much less diverse and much, much smaller than in the past.

NATO's tactical nuclear forces were in the past. Of course, if the United States were to actually try to remove nuclear weapons from European storage sites with the idea of really using them, the host countries' reactions might be very different. Still, if the honeymoon with Russia is really over, the United States is going to have to think carefully about its nuclear commitment to NATO, and NATO is going to have to reevaluate its future role.

No other conventional imbalance comparable to the old Cold War NATO/Warsaw Pact situation exists now or is likely to exist *in general* for the foreseeable future.[20] Indeed, it is unlikely that any other country would even try to match U.S. conventional capability for some time to come. The most likely danger is of a different sort. First, U.S. dominance in all aspects of power—nuclear and conventional military, economic, even cultural—will inevitably make other nations, even traditional allies, uneasy and is likely to provoke reactions of various sorts.[21] Alliances are likely to readjust and form accordingly. For example, one obvious possibility is for newly elected Russian president Vladimir Putin to try to play a "China card" at U.S. expense. Readjustments of regional alliances are likely as well. The net effect will probably be an array of subtle—and, perhaps, not-so-subtle—alterations in the world political scene, which U.S. foreign policy will have to be sufficiently nuanced to deal with.

Second, at the military level, others may not even try to match the United States in specific military capabilities. Rather, they will react by trying to counter particular U.S. strengths in other ways. An obvious example is the Russian declaratory policy of increasing its emphasis on tactical use of nuclear weapons, including first use, to compensate for its conventional weakness (and by implication, U.S. conventional superiority). Others, who have no reason to accept "rules of the game" that are largely shaped by U.S. preferences and

---

[20]One of the few, albeit remote, possibilities is a future world in which China, with more mobile armed forces, decides to invade, say, Siberia for some bizarre reason, and the United States—for some equally bizarre reason—decides that it cares enough about the invasion to try to stop it militarily.

[21]See, for example, Marshall and Mann (2000) and McManus (2000) for discussions of worldwide reactions to U.S. preeminence.

legitimize and reinforce U.S. power, have gone further. The United States will have to be ready to adjust to such policies as they emerge.

However, local and temporal conventional imbalances can still occur. With U.S. presence overseas continuing to shrink, it will take some time to move enough U.S. forces to a distant theater to fight a major campaign. For example, during Operation Desert Shield, it took months to move to the Persian Gulf all the allied forces that were eventually used in Operation Desert Storm. During that time, the enemy was passive, not attempting to improve its position by taking advantage of its fleeting conventional superiority or trying to interfere with U.S. force deployments. The next enemy, having learned the lessons of the Gulf War, may not be so accommodating. In cases where important U.S. interests are at stake, the United States could face some difficult choices. Massing conventional firepower at long range is difficult and expensive, particularly if local bases are unavailable because of political considerations or enemy capabilities (e.g., nuclear or chemical threats to airfields or ports). The burden would normally fall on the U.S. Air Force and the Navy, if the local geography permitted, to provide enough high-quality firepower to halt an invasion or deal with whatever problems the theater conflict posed, at least long enough for other forces to be deployed to the theater.

Those logistics could prove increasingly difficult against competent opposition if U.S. forces operate in traditional ways. Long-range operations stretch the bomber force, which has been shrinking. Both the Air Force and the Navy have been emphasizing relatively short-range strike aircraft, which may be inappropriate for future regional conflicts. Even worse, the United States may not have enough of the right kinds of quality conventional weapons. As the operations in Kosovo demonstrated, the United States does not have a large enough inventory of precision-guided weapons in general. Particular classes of accurate weapons are in even shorter supply. The Air Force has invested heavily in short-range weapons, which would be appropriate for a bomber force that included a large number of B-2s. However, because the Air Force procured only a small B-2 force, it needs more long-range weapons to make its older bombers effective in a hostile air defense environment. It has not done that. As a result, *the United States could face real challenges in a stressing future conventional campaign against a quality opponent* unless it restructures its conventional forces.

Could U.S. nuclear weapons help fill this gap? That depends on a number of factors. First, the situation has to be appropriate. If nuclear weapons are to be used to win the war, there have to be suitable targets—e.g., massed forces, airfields. There have to be effective operational options and adequate forces available for the United States to use nuclear weapons effectively. More fundamentally, to justify serious consideration of using nuclear weapons:

- The immediate stakes in the conflict have to be high enough.
- The collateral damage has to be "acceptable," whatever that means in context.
- The short-term gain has to be adequate to justify the long-term costs of using nuclear weapons.

This is a difficult set of criteria to meet, particularly given the United States' dominant military position in the world and the relative absence of potential overseas conflicts that in fact affect its vital interests.

There is always the possibility of merely threatening to use nuclear weapons to punish a regional adversary if it does not stop doing whatever it is doing. The targeting and operational issues become easier, but the political criteria are, if anything, more demanding. In addition to meeting the criteria above, nuclear threats would have to be credible to the potential victim. That has been devilishly hard in practice over the years, particularly if the quarrel at hand was more important to the regional players than to the intervening superpower, and there was no history of superpower interest in the region. An example was the almost quaint attempt by Boris Yeltsin to rattle Russian nuclear sabers over NATO's action in Kosovo and its exclusion of Russia from the decisionmaking process. In spite of Russia's nuclear weapons and its historic affinity for Yugoslavia and the Serbs, everyone ignored Russia's nuclear threats, as they should have.[22]

[22]Of course, there is always the unsettling question: What if Yeltsin had been serious? Because Russia made the nuclear threat from a position of extreme political, economic, and conventional military weakness, all of which made Yeltsin's threat less

Ironically, Russia's surprise move of troops into Kosovo had much greater impact on NATO than its incredible nuclear threats.

The United States could encounter threats from other than conventional or nuclear weapons. The possibilities receiving the most attention are chemical and biological weapons (CBWs). Both have actually been used in warfare and can be more lethal than conventional weapons, depending on the circumstances. Both are stigmatized by the international community.

Biological weapons in particular appear to have become the weapons du jour for nations or, perhaps more likely, for transnational terrorist groups trying to cause large-scale death. Although they have many practical disadvantages as weapons, they have a number of potentially attractive features for nations or groups that cannot build or afford either nuclear weapons or large and modern enough conventional arsenals to directly engage a major military power such as the United States:

- They are relatively easy and cheap to develop, especially compared with nuclear weapons. (Unlike nuclear weapons, at least some kinds of potentially lethal biological agents are within the capabilities of virtually any journeyman microbiologist.)
- Development and production facilities are easy to disguise.
- The agents themselves are easy to deliver covertly, although dispersing the agents effectively and efficiently could be very difficult.
- The origin of the attack might be very difficult to determine.
- The level of damage *might* be quite substantial relative to the effort required to develop and use the weapons

The use of biological agents in particular might provide either nation states or terrorist groups with the means to strike directly at the U.S. homeland. Regional powers or smaller nations might find such

---

credible in the absence of a clear and substantial Russian stake in the outcome of the conflict. Still, that might have actually made a Russian leader more desperate, which could have immensely upped the stakes for everyone involved in the conflict. That is precisely the danger of nuclear brinkmanship. Even under the best of conditions, sending and receiving clear signals can be very difficult.

weapons the most effective way to raise the cost to the United States of a war that they were losing. It might be their best hope of improving the outcome of any settlement of a war, particularly if they were already desperate enough to be indifferent to further conventional or even nuclear attacks.

CBWs used on the battlefield can hamper military operations by forcing troops to wear protective gear, take inoculations, etc. Use of such weapons could complicate operations at airfields and ports, requiring aircraft and ships exposed to chemical or biological agents to be decontaminated. Without adequate advance preparations, chemical or biological weapons might be much more lethal to personnel than most conventional weapon attacks.

In addition to passive defense against CBW attacks, threats of greater violence—either nuclear or conventional—in response to CBW use may be sufficient to deter that use if the enemy can still be deterred. Iraq was apparently deterred from using chemical weapons in the Gulf War by an unambiguous U.S. threat to use nuclear weapons in response. Nuclear weapons generally trump either biological or chemical weapons in either threat or application. Thus, invoking nuclear weapons as a threat to deter CBW use is a logical option and is now explicitly stated as U.S. policy, although the details are a little vague.[23]

Beyond their deterrent value, nuclear weapons might be used directly against biological and chemical weapons facilities to defeat an enemy's capability to produce or stockpile such weapons, assuming, of course, that the facilities can be identified and characterized properly. This kind of nuclear targeting has been explored in great detail in recent years. One of the issues is whether nuclear weapons are appropriate, necessary, sufficient, or credible. The next section and the companion volume to this report examine some of the analytical comparisons between using nuclear and conventional weapons for such attacks. United States Strategic Command (U.S. STRATCOM) and others make these calculations routinely, so there is no great mystery there. Still, the relative effectiveness of nuclear weapons over other options for various kinds of tasks is central to the larger

[23]Warner (1999).

questions of what role nuclear weapons ought to play in future U.S. security policy and what sorts of forces are needed to support that policy.

The idea of using either nuclear or conventional weapons to deter or counter CBW embodies one of the central themes that we have identified in analyzing contemporary nuclear strategy: *the importance of asymmetry.* Fortunately, the United States explicitly recognizes that it does not require chemical and biological weapons of its own to respond to CBW threats from others. Different kinds of responses could be more effective, more appropriate, and more in synch with the "American way of war." We will develop this theme subsequently as we examine future U.S. strategic options.

On the other hand, there is an unfortunate countertrend that tends to lump nuclear weapons and CBWs under a common rubric of WMDs. Aside from the fact that chemical weapons do not really belong in the same category as the other two, there is a more fundamental problem. Presumably, the reason for including CBWs in the same category as nuclear weapons is to help lay the political and cultural groundwork for a U.S. policy of threatening nuclear use against CBW facilities of others. The implicit assumption appears to be that only "response in kind" (e.g., tit-for-tat, nuclear threat begets nuclear response) is acceptable. Imposing such symmetry on U.S. strategic options could narrow U.S. policy choices more than necessary. Asymmetry cuts both ways.

Striking what may be a resonant chord for future U.S. adversaries, two Chinese military officers have expanded the asymmetry concept still further.[24] They argue that China should take a much broader and more nuanced approach to developing a military strategy to fight a stronger, richer military power such as the United States. Approaches could include terrorism, drug trafficking, environmental degradation, and computer hacking in addition to more traditional military forces. Interestingly, they see theater ballistic missile defense as a U.S. ploy to lure China into a traditional arms race that it would surely lose. (Ironically, we think China would win an arms race that pitted their ballistic missiles against U.S. ballistic missile

---

[24]Pomfret (1999).

defenses.) Instead, they proposed a different strategy entirely. *The United States should expect more countries to adopt this approach and look for different ways to attack the United States and its interests.* The United States, in turn, needs to develop a strategy to take maximum advantage of what it does well.

To sum up, the United States may face a much more diverse set of threats in the future as weaker countries look for ways to offset the U.S. advantages in traditional military forces. Some of these could permit direct attacks on the United States itself. Others could raise the cost of U.S. military actions in regional conflicts. Threats of retaliation may deter some of these actions; in other cases, they may not. The problem for the United States is to develop its own security strategy and the appropriate forces to execute that strategy.

U.S. NATIONAL SECURITY POLICY: A SPECTRUM OF POSSIBILITIES

Figure 3.1 shows a hierarchy of approaches that the United States could take in dealing with future conflicts. They are ranked more or less in order of preference, all other things being equal, if they were feasible to implement.

Abstinence

Although the U.S. military does not have the luxury of choosing its quarrels, there is a strong a priori argument that the United States has the option of being much more selective in using its military forces in the current world than it did in the past. Any great power in the history of the world would envy the current U.S. national security situation. With the end of the Cold War, the United States faces few direct threats to its security and has few interests in the world that are critical enough to be worth fighting over. Few of those interests that are worth fighting for involve nuclear powers, especially with the reduced chances of a superpower collision in local quarrels.[25]

---

[25]Exceptions exist, of course, and tend to prove the rule. Some regional disputes potentially involving nuclear powers still remain as artifacts of the Cold War (e.g., North Korea-South Korea, China-Taiwan). More recently in NATO's Kosovo opera-

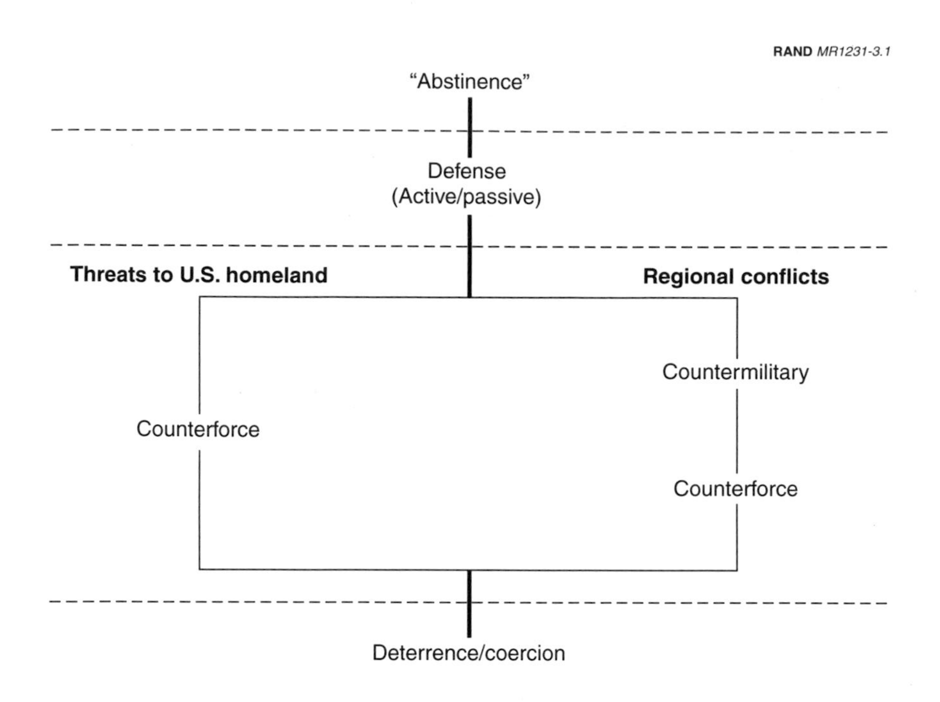

Figure 3.1—Hierarchy of Approaches to Dealing with Future Conflicts

Moreover, given the relative military and economic power of the United States, even Russia no longer qualifies as a superpower in spite of the fact that it could still destroy the United States with a nuclear attack.

On the other hand, all conflict involves some inherent risk, particularly in a world where terrorist attacks or other unconventional threats could affect the U.S. homeland directly. Thus, avoiding unnecessary involvement in peripheral conflicts has a lot to recommend it as a national policy.

---

tions, the United States managed to alienate both Russia and China, the only two current nuclear powers likely to pose a nuclear threat to it.

When conflicts do occur, there are generic options for dealing with them. Figure 3.1 shows them in descending order of preference, although in practice some mix can be expected.

Defense

If defenses could be made to work adequately, they could form the basis of an effective military and political strategy because they are just that: defensive. They only protect; they do not attack. Therefore, they can be used only in response to an attack by an aggressor and generally do minimal damage to others. Who could argue with such a strategy if the systems were there to make it effective? It captures the moral high ground and avoids most of the problems that more aggressive strategies have in worrying about damage to surrounding areas, command and control, and the like. That, of course, was the basic rationale for the Strategic Defense Initiative (SDI) of the Reagan era, although the technology to support the concept did not exist.

As noted earlier, passive defenses can be an important part of an overall capability to cope with chemical or biological weapons. Against nuclear weapons, passive defenses are of limited value and tend to be expensive.

As a practical matter, effective active defenses are difficult to develop. The United States has effective defensive systems to protect against aircraft. However, it has never deployed a North American air defense that could protect its borders adequately against lone intruders, much less against a large-scale air attack.

Cruise missile defense is an extension of air defense. It is generally expensive because of the density of defenses typically required and is technically demanding in some areas. Still, this is a problem the United States knows how to solve.

Ballistic missile defense is another matter. In spite of a small number of recent tests that finally achieved successful intercepts in benign environments, vast uncertainties remain:

- Critical technology remains unproved
- An effective system is potentially very expensive
- Operational issues abound.

As we have pointed out in earlier work, determining the feasibility of ballistic missile defense and deciding whether to "bet the farm" on it is probably the most difficult decision that the United States—and the Air Force—will have to make in the nuclear arena.[26] It will affect fundamental U.S. nuclear strategy and choices regarding nuclear weapons in the most fundamental way. Defenses that worked well enough could allow, for the first time, a defense-dominant nuclear strategy.

Counterforce and Countermilitary Operations

Absent leakproof defenses, the next option is a range of attacks on enemy forces. There are at least two general classes of relevant military operations. In theater conflicts, a broad range of attacks on enemy military forces is routine; it requires a decision to initiate attacks by the United States and its allies, but in the face of action by an enemy—especially launching an invasion—such a decision should not be particularly demanding.

Counterforce attacks—attacks on an enemy's nuclear forces, as that term is generally used in this context—are a much more serious matter. They involve attacks, perhaps with nuclear weapons, on an enemy's homeland. To be fully effective, they require striking first. During the Cold War, the concern with such attacks was that they might not be effective and might indeed precipitate precisely the effect that they were intended to prevent (i.e., a nuclear attack on the United States). Moreover, pursuit of such capabilities fueled a strategic arms race that was not only expensive but arguably left the United States worse off.

Much has changed over the years in the technical feasibility of launching counterforce attacks. Nuclear weapons systems are much more effective than they once were. So are conventional weapons.

---

[26]Glenn Buchan, *Nuclear Weapons and the Future of Air Power*, RAND (forthcoming).

Revisiting counterforce is a fundamental element of any future nuclear strategy.

Deterrence and Coercion

All the options considered so far—everything above the lowest dashed line in Figure 3.1—have an important feature in common: The outcome rests with decisions that the United States makes. For example, faced with a nuclear threat, if the United States had the capability, through some combination of counterforce capabilities to destroy enemy weapons before they were launched and active defenses to defeat any weapons that were launched, it could protect itself no matter what an enemy chose to do.

Deterrence and coercion are fundamentally different. They depend on influencing the decisions of others. In principle, that is clearly less desirable because in the case of nuclear threats to the United States, for example, it places U.S. survival in the hands of potential enemies. Similarly, enemies get to make final decisions about lesser actions that the United States might seek to influence.

Deterrence by threat of retaliation has fundamental conceptual weaknesses. It remains problematic what deters whom from doing what to whom. Empirical evidence for the effectiveness of deterrence is limited and ambiguous.[27] Although *capability* itself may be sufficient to deter, the credibility of threats is always a potential issue. Opponents may simply not *believe* nuclear threats in particular.

Deterrence by threat of nuclear retaliation was the centerpiece of U.S. strategy for dealing with the threat posed by the massive Russian nuclear arsenal during the Cold War. There was no way to be certain that it would work, but it was better than any of the alternatives. The issue for contemporary U.S. military policy is where deterrence fits in the spectrum of U.S. options, and what instruments are most likely to be effective in implementing that policy.

When we refer to "deterrence" here, we specifically mean deterrence by threat of punishment, nuclear or otherwise. That is the context in

---

[27]See Payne (1996), particularly pp. 97–116.

which deterrence acquired special connotations in the nuclear arena. Unfortunately, as soon as the concept of "nuclear deterrence" became politically acceptable, every application of nuclear weapons was recast as providing some form of "deterrence" (e.g., "deterrence by denial," which is logically indistinguishable from actual war-fighting capability), since deterrence sounded benign. Now, there is nothing logically wrong with that. *All* weapons are supposed to frighten adversaries and give them pause before starting a conflict that they are likely to *lose.* However, *lumping all applications of nuclear weapons under the general rubric of "deterrence" devalues the concept.* Jaded audiences hearing continuing rationalizations of nuclear forces of all stripes as necessary for "deterrence" can perhaps be forgiven for being confused and a bit cynical. That is why we try to use more precise language and say what we mean.

POTENTIAL ROLES FOR U.S. NUCLEAR WEAPONS

Every U.S. administration has a stated policy on nuclear weapons. For example, speaking for the Clinton administration, then-Assistant Secretary of Defense Edward L. Warner III laid its policy out for Congress.[28] Some of the key points that former Secretary Warner made were the following:

- Nuclear weapons remain a vital part of U.S. national security policy.
- The United States cannot eliminate nuclear weapons or even reduce forces to low levels (e.g., a few hundred weapons) for the following reasons:
  - The United States needs a hedge against a resurgent Russian nuclear threat.
  - The United States needs to deter China.
  - The United States needs to deter rogue nations that develop weapons of mass destruction and long-range delivery systems.

---

[28]Warner (1999).

While this statement is not bad by the standards of policy pronouncements produced by the bureaucracy and even has some interesting elements (e.g., the explicit expectation that U.S. nuclear weapons might help deter rogue states from using chemical or biological or chemical weapons), it is at once too vague and too specific to make a compelling case for what kind of nuclear capabilities the United States needs in the current world and why. It is too vague because it does not explain what explicit strategy we would rely on to deal with, for example, a revanchist Russia. It is too specific because it appears to limit the role of U.S. nuclear weapons in nonnuclear situations to deterring use of biological or chemical weapons.

The United States needs to think through this issue more carefully. Accordingly, we have taken a broader look at potential future roles of U.S. nuclear forces.

Figure 3.2 shows a range of possible uses for U.S. nuclear forces. The United States has considerable latitude in choosing what it would like its nuclear forces to do in the contemporary world. The choices will be strongly affected by a variety of factors, some of which are shown in the figure. Generic classes of missions are shown along the horizontal axis, along with a brief summary of critical factors for each class of mission. The vertical axis is a subjective assessment of the relative importance of these generic classes of possible missions to U.S. national security (e.g., deterring by threat of retaliation, particularly of nuclear or other massive attacks on the United States, is primary; reducing damage to the United States should deterrence fail is next; other missions are less significant). The shading on the figure suggests the relative importance of nuclear weapons in each role.

Terror Weapons/Traditional Deterrence

Nuclear weapons are unmatched as terror weapons and are therefore the most effective possible weapons to implement a policy of deterrence by threat of punishment. This is the most enduring role for nuclear weapons and the one for which they are most uniquely suited. The only issue is whether the United States wants to continue to have this kind of capability and whether it needs to inflict the levels of damage that nuclear weapons can cause.

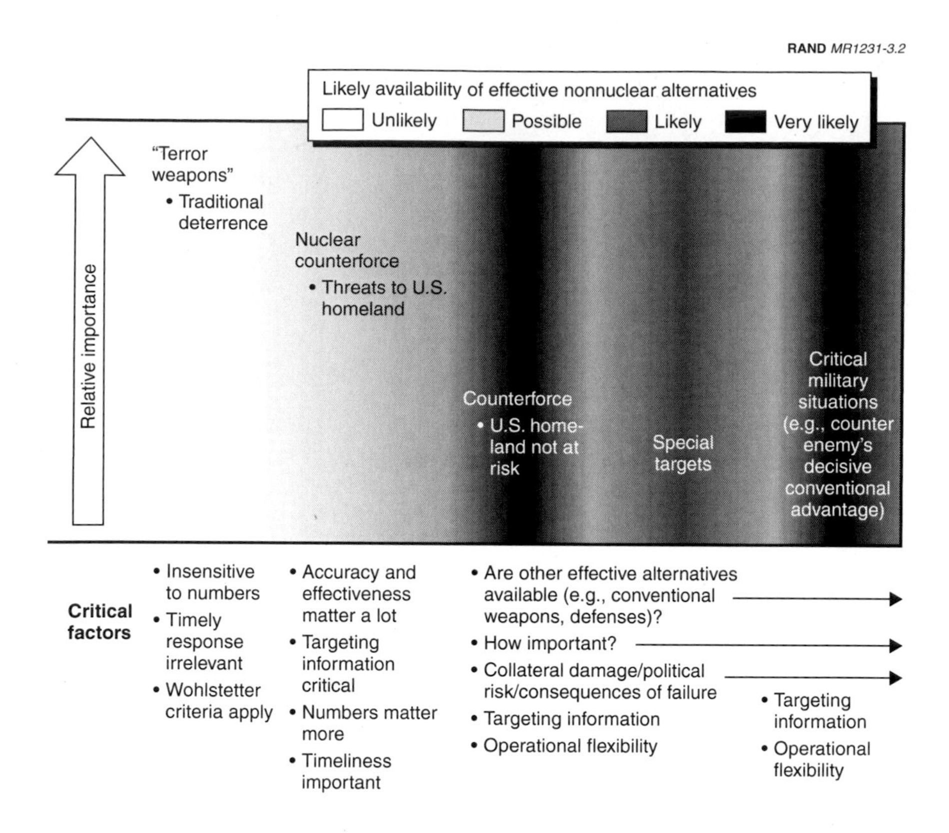

Figure 3.2—Why the United States Might Want Nuclear Weapons in the Contemporary World

Forces do not have to be particularly large to perform this role. Indeed, "requirements" for levels of punishment needed to deter are largely arbitrary. Thus, force levels are likely to be determined more by practical considerations such as cost. However, nuclear forces and their associated command and control systems do need to be survivable against competent attacks. They also must be operated safely and securely.

Interestingly, timely use of nuclear weapons is *not* required. Response has to be certain and properly directed, but *not* swift.

Counterforce[29]

The other role for which nuclear weapons are especially well-suited is counterforce, particularly counterforce attacks against enemy weapons that can reach the United States. With the weapon accuracy improvements over the last couple of decades, debates about the potential effectiveness of nuclear weapons to destroy hardened fixed targets such as missile silos have long since been resolved. Mobile targets remain elusive, but if adequate surveillance capability can ever be achieved, destroying them with even conventional weapons should be relatively easy. In fact, modern conventional weapons might be quite effective against hardened fixed targets. However, if enemy nuclear forces could be targeted against the U.S. homeland, *the stakes are probably high enough to use the most effective weapons available—nuclear weapons.*

Details of weapons performance and force structure are much more important for a counterforce strategy than for deterrence by threat of punishment. Timeliness is vital. Requirements are much more real.

An irony of the end of the Cold War is that *the sort of counterforce strategies that the United States planned for during the Cold War—and that had no chance of being effective once the Soviets developed hardened silos and missile-launching submarines (SSBNs)—might in*

---

[29]We use "counterforce" to mean the targeting of enemy nuclear forces to limit damage to the United States in a nuclear exchange. That is the way the counterforce concept was originally developed at RAND in the 1950s (see the discussion in Kaplan or the original internal memoranda by our colleague, James Digby, who directed most of the early counterforce work at RAND). This usage became even more explicit as targeting concepts evolved and the "counterforce" jargon was associated explicitly and exclusively with attacks on enemy strategic nuclear forces, as opposed to "countermilitary" targeting, which covers the entire spectrum of military targets; targeting "war supporting industries," which was supposed to reduce an enemy's long-term capability to make war (and in later years sounded more palatable politically than economic targeting in general and more feasible than targeting to inhibit an enemy's ability to recover economically after a general nuclear war); and "leadership" targeting, which is self-explanatory, at least in principle. This jargon has been used—and these distinctions recognized—essentially universally in the U.S. nuclear community for decades. That said, the expression "counterforce" like the expression "deterrence," has been misused and abused occasionally, generally by those who want to blur distinctions in targeting philosophies. Since targeting is the most direct manifestation of strategy, and identifying and evaluating alternate nuclear strategies is the objective of this study, we obviously are obliged to sharpen rather than blur such distinctions, even granting some inherent ambiguity.

fact work in the current world, particularly against fledgling nuclear powers that have not yet learned how to play the game. Even the Russians theoretically could be vulnerable to a U.S. counterforce attack if they kept their SSBNs in port and mobile ICBMs in garrison. Absent adequate tactical warning, their strategic forces would be potentially vulnerable to attack by a relatively large, high-quality nuclear force. Some U.S. analysts have worried considerably about this potential Russian vulnerability.[30]

However, a U.S. nuclear attack against Russian nuclear forces would be nearly unthinkable in the current world under any but the most dire circumstances (e.g., a fragmenting Russia, if Russian nuclear forces fell into the hands of psychotics). When the United States and Russia were drastically reducing their nuclear forces and alert levels in the wake of the Cold War, both sides—but particularly Russia—understood that they were increasing their vulnerability to a nuclear first strike. However, in the political climate of the times, *they did not care.* A fundamental question that the United States has to ask now is whether it still believes that the political climate is sufficiently benign that relaxing the old Cold War operational practices is still warranted.

The United States has a relatively straightforward choice to make: how much counterforce capability does it want to include in its future nuclear force posture? Even small U.S. nuclear forces would have considerable inherent counterforce capability against emerging nuclear powers that had only small numbers of somewhat vulnerable nuclear weapons. Thus, any nuclear force the United States is likely to deploy would be capable of meting out substantial punishment.

The same is *not* true of U.S. forces designed to launch counterforce strikes against Russia. U.S. nuclear forces would have to be much larger than they would if simple deterrence or counterforce against minor nuclear powers were their only objectives. Others will be able to discern those differences as well. *"Selling" such a U.S. force as merely a "deterrent" will be virtually impossible.*

---

[30]See, for example, Blair (1995) pp. 71–72, and Blair's chapter in Feiveson (1999), especially pp. 109–111.

Even if the United States chose to pursue and extend its current counterforce advantage, such an advantage is likely to be fleeting. The counters are well known—they just take time, resources, and experience to implement. In any case, *deciding how robust a nuclear counterforce capability the United States wants to maintain will be a major determinant of how far the United States is willing to reduce its nuclear forces and how it chooses to operate them.*

Interestingly, active defense (e.g., ballistic missile defense [BMD]) could serve the same function as counterforce. Nuclear warheads could play a role there, too, although current U.S. ballistic missile and air defense concepts rely on conventional warheads.

Special Targets

In recent years, there has been even more interest than in the past in attacking special kinds of targets that are very difficult to destroy. Underground facilities, especially deeply buried targets, have received considerable attention in this regard. Such facilities could include command centers, manufacturing plants or storage sites for special weapons, or other types of high-value installations. Part of the argument for maintaining a capability to attack such targets, even if nuclear weapons are necessary, is the most fundamental tenet of deterrence: Deterrence requires holding at risk whatever the enemy values. If the enemy cares enough about a facility to bury it deeply underground, then that may make it worth being able to attack on first principles.

Obviously, there is more to it than that. For example, an enemy may value an installation that the United States does not want to destroy. A classic example is command and control. Attacking command and control targets can either be a very good or a very bad idea, depending on the situation and the attacker's war aims. It is conceivable that the United States might conclude that leaving an enemy's command and control system intact would be in both sides' interest.

In evaluating the feasibility and desirability of nuclear weapons against these types of targets, there are a number of questions that need to be answered:

- How important are these targets to U.S. objectives?
- Are they important at all
  - Are they worth the cost of an attack?
  - Under what conditions would they be important enough to warrant a nuclear attack?
- Must they be destroyed to produce the desired effect, or are there other options for achieving a "functional kill" (e.g., destroying communications antennas, sealing entrances)?
- Is destroying the targets feasible, even with nuclear weapons?
- What sort of nuclear weapons would be appropriate for this type of application?[31]
- How do nuclear weapons and other options compare in terms of
  - likely effectiveness?
  - technical and operational difficulty?
  - number and type of weapons required?
  - collateral damage?
  - long-term political costs?

Some of these are analytical issues. Others are subjective. Chapter Four includes preliminary analysis of weapons effectiveness in selected cases to help narrow the list of practical choices.

Critical Military Situations

Finally, there is a whole set of military situations that could arise in which nuclear use might be an option if the situation were important enough and other options appeared inadequate. During the Cold War, unfavorable conventional balances in Europe and, to a lesser degree, in Korea drove the United States to develop tactical options for nuclear use to offset its conventional deficiencies. Since then,

[31]The United States already has in its inventory the B61-11 earth-penetrating nuclear gravity bomb, specifically designed to attack underground targets. The B-2 bomber is the delivery platform of choice.

conventional weapons have improved considerably, particularly U.S. conventional weapons. There are fewer potential situations in which the United States would have a compelling stake in the outcome of a conflict and would be at a serious conventional disadvantage. Still, questions remain:

- Are there any contemporary situations that are important enough and difficult enough to warrant the threat or actual use of U.S. nuclear weapons?
- If so, what sort of nuclear forces and operational procedures would be necessary or appropriate for dealing with these contingencies?
- How much more effective than conventional weapons would nuclear weapons be, and are the differences worth the cost (e.g., collateral damage, political costs)?

Again, the answers to some of these questions turn on analytical issues, specifically the relative effectiveness of modern conventional and nuclear weapons. The next chapter compares conventional and nuclear weapons in at least one potentially stressing battlefield application. Such comparisons are critical to deciding under what conditions the nuclear option is worth pursuing.

The relative importance of developing actual war-fighting nuclear options is the critical element in deciding on what future course U.S. strategy ought to take. The problem bifurcates into using nuclear weapons only to deter others by threatening punishment or also using nuclear weapons as war-fighting instruments. The first is familiar and straightforward. The second includes a richer set of possibilities and has different implications for forces and operational procedures. Chapter Five will address some of those implications. All of these capabilities are achievable to some degree, and they lead to significant variations in future U.S. nuclear strategy. Thus, choices abound.

Chapter Four

STRESSING CASES: SOME CONTEMPORARY COMPARISONS BETWEEN NUCLEAR AND CONVENTIONAL WEAPONS

Primary roles of the U.S. nuclear deterrent force are to defend the United States against a large strategic attack and to be capable of projecting enormous damage against an adversary. These roles may not be the only situations in which nuclear weapons would be used. Indeed, for decades, the United States has included in its inventory nuclear weapons that were actually intended for battlefield use and has developed its war plan with that in mind. Use of nuclear weapons in combat to solve actual military problems may meet concepts of proportionality in conflict more clearly than use of nuclear weapons in retaliation. That is, the use of a nuclear weapon to solve a given military problem (such as we will discuss below) will probably result in fewer casualties and collateral damage than a retaliatory use of nuclear weapons against, for example, population centers (whether population centers are targeted deliberately or inadvertently). The latter might fail a proportionality test, and even threats of nuclear use might not be credible for that reason. Of course, even relatively "clean" tactical use of nuclear weapons could fail a proportionality test if it triggered escalation that led to larger-scale, less-discriminate nuclear use. During the Cold War (and still today, in theory), NATO's nuclear strategy depended on the threat of precisely such an escalation process to deter the Soviets from attacking western Europe. Conversely, the fear of just this kind of escalation or even a longer-term alteration of the nuclear "rules of the game" that would encourage nuclear use is the strongest argument against even effective, limited use of nuclear weapons to solve tactical problems.

With the demise of U.S. tactical nuclear forces, strategic forces might be the only option available to deal with military situations that conventional forces cannot handle. To illustrate the relative effectiveness of nuclear and modern conventional weapons, we examined several selected cases that might stress conventional forces. Four classes of scenario are considered in this chapter: (1) halt invading armies, (2) destroy hardened bunkers containing WMD, (3) destroy a deeply buried facility, and (4) defense against ballistic missiles.

STRATEGIC NUCLEAR WEAPONS IN HALTING INVADING ARMIES

The first application of nuclear weapons that we will examine is the use of nuclear weapons for halting an invading army. We will consider a parametric range of possible warhead yields (1, 10, 100, and 1000 kT). As we will show, for many potential applications nuclear weapons in the 1–10 kT (or even subkiloton) range would be sufficient. Conversely, we will also show that there are some cases where high yield (~1 MT) weapons are the only weapons capable of assured destruction of particularly difficult targets.

During the Cold War, using nuclear weapons to halt a possible Warsaw Pact invasion of Western Europe was a staple of NATO planning. However, recent developments in conventional weapon technology have made it possible for a relatively small amount of air power to halt an invading army without resorting to nuclear weapons. This advanced technology and the capabilities it provides have accentuated the tradeoffs between conventional and nuclear weapons use in this scenario. These tradeoffs need to be considered even in a scenario whose urgency would previously have demanded the use of nuclear weapons.

As a first step toward examining these tradeoffs, we briefly review the capability of both nuclear weapons and current smart/brilliant weapons (e.g., Sensor-Fuzed Weapons [SFWs] and Brilliant Anti-Tank [BAT] munitions) to halt an invading army. Then we examine the pros and cons of nuclear weapons versus smart weapons in halting an invading army. We compare the effectiveness, platform requirements and limitations, and the collateral damage probability of both nuclear and smart weapons.

Halting an Army: Nuclear Weapons

To review the capability of nuclear weapons to halt an invading army, we used effectiveness values calculated from the *Physical Vulnerability Handbook for Nuclear Weapons* (PVH).[1] We calculated values for damage against tanks in a warned/protected troop posture. We also compared weapon effectiveness in two cases: a height of burst (HOB) selected to maximize the area affected ($HOB_{optimum}$), and a HOB high enough above ground to avoid appreciable fallout. The HOBs were selected from those listed in the PVH. $HOB_{optimum}$ is the HOB that gives the maximum weapon radius (WR) for a given yield. (WR is equivalent to the radius of weapon effects.) The minimum HOB to avoid appreciable fallout is $HOB_{min} \sim 54 \cdot W^{0.4}$ meters, where W is the weapon yield in kilotons[1]. We always selected the larger of the two heights of burst. If $HOB_{optimum}$ exceeded HOB_{min}, we used $HOB_{optimum}$ and got maximum effectiveness with no fallout. If HOB_{min} exceeded $HOB_{optimum}$, we used HOB_{min}, sacrificing some effectiveness to reduce collateral damage from fallout.

Choosing a damage level of immediate permanent ineffectiveness (the most severe level—see below), we found values of WR.[2]

The weapon radius given in the PVH is calculated using both radiation and airblast rather than one or the other as the dominant weapon effect. The procedure involves calculating the effects independently, ignoring the possible greater susceptibility of a target to one effect because of exposure to another. Note that although the PVH considered the effects of airblast in damaging tanks, it is primarily the radiation that results in fatal casualties among the tank crews.

Assuming a negligible circular error probable (CEP), the WRs were used to calculate the probability that the target (in this case, a tank) will be damaged to the level specified (i.e., immediate permanent ineffectiveness) at a given distance from ground zero. These prob-

---

[1]None of the calculations or results from the PVH shown in this paper are classified.

[2]David Matonick and Calvin Shipbaugh, *Selected Comparisons of Nuclear and Conventional Weapons Performance*, RAND (forthcoming). Government publication; not releasable to the general public.

abilities as a function of distance from ground zero are shown in curves contained in the companion report.

Lower-yield nuclear weapons are probably more attractive than higher-yield weapons for halting armor, as is consistent with U.S. plans during the Cold War to use tactical nuclear weapons of much smaller yield-to-target company-size armored units to counter a Warsaw Pact invasion of western Europe. This attractiveness is a result of the radiation effects (even without an enhanced radiation warhead) in the low-yield weapon, which was adequately effective against armor with much less collateral damage than higher-yield weapons would have caused.

Halting an Army: Smart/Brilliant Weapons

The capability of current smart/brilliant weapons to halt an invading army has been examined in detail elsewhere. Our review will compare smart/brilliant weapon capabilities with the capabilities of nuclear weapons. First, we will describe the various nuclear damage levels used in the PVH for a troop posture of warned/protected tanks.

We have been using the most lethal of the damage levels—immediate permanent ineffectiveness. The PVH also defines two others: immediate transient ineffectiveness and latent lethality. In addition to the damage criteria for airblast and radiation, we also calculated the weapon radius and R_{50} for a 100-kT warhead detonated at HOB = HOB_{avoid}. The value R_{50} is the range at which there is a 50 percent probability that the target will be damaged to the level specified. Using the values of R_{50}, we made a simple comparison of nuclear effects radii with those of SFW.

Field tests have shown that 47 percent of the tanks contained in an SFW footprint were "availability killed" ("A-killed"). That is, at least one critical component was damaged, forcing the tank to leave the line of march.

Two overlapping SFW footprint regions give an A-kill probability of $P_k = 1 - (1 - 0.47)^2 = 0.72$. Thus, the overlapping SFW footprints from one sortie have an A-kill probability of 72 percent. Note also that a single F-16 can carry four SFWs.

Therefore, approximately 2400 meters of road can be serviced to the A-kill standard (with a P_k of 72 percent) by four F-16s carrying SFW, provided the SFW can be accurately placed along the road (which is becoming more likely with the introduction of new weapons—e.g., Joint Standoff Weapons [JSOWs] and Wind Corrected Munition Dispensers [WCMDs]).

A basic assumption in previous RAND research has been that enemy units advance until about 50–70 percent of their armored vehicles have been damaged to at least the A-kill standard. Therefore, a sufficient number of SFWs (we give an example of how many below) placed accurately along a road (for the case of armor units in a road march formation) will be quite capable of halting an advancing armor unit. We will look at this capability again below in considering platforms.

Halting an Army: Collateral Damage Possibilities

One important point to note with BAT or SFW is the low probability of collateral damage or fratricide associated with these weapons, as compared with nuclear weapons. BAT uses acoustic sensors initially to locate a target cluster, and then relies on an infrared (IR) sensor to terminally guide itself to an individual vehicle, which it then destroys with an armor-penetrating warhead. SFW munitions operate in much the same way as the terminal phase of BAT. The SFW round (called a skeet) fires an explosively formed armor-penetrating slug at the IR-bright region of a hot vehicle engine.

In addition to being accurate, BAT and SFW are designed mainly to destroy vehicles by penetrating the engine block (guiding in on the strong IR signal located there) and producing an A-kill. Therefore, it is (albeit remotely) possible for the vehicle crew to survive an attack by these weapons (even though their vehicle is killed). Any dismounted friendly units are unlikely to suffer serious casualties. (Recall the Gulf War experience of Iraqi tank crews who found it safer to sleep *outside* their tanks.) Collateral damage will likewise probably be low or at least much lower than if a nuclear weapon were used to halt an invading army.

Clearly, the nuclear weapon damage levels used above are far in excess of the A-kill standard, even for the latent lethality damage level.

This fact raises the question of whether nuclear weapons are more destructive than necessary for halting an invading army. To address this question, we must consider the thermal radiation effects of nuclear weapons. Detonating a nuclear weapon at a high enough altitude to avoid fallout will in fact increase some collateral damage problems. Thermal effects of nuclear weapons will increase as they are detonated at higher altitudes (for example, up to about 3000 feet for a 100-kT warhead).

To examine this type of collateral damage effect, we calculated curves of thermal radiation as a function of ground range using a nuclear weapon effects calculator in which we could adjust both yield and height of burst. Curves for 1, 10, 100, and 1000 kT are shown in Figures 4.1 to 4.4. HOBs selected were ground burst, HOB_{min}, and $HOB_{optimum}$. Also shown are the approximate thermal radiation levels for first- and third-degree burns, and for igniting wood.

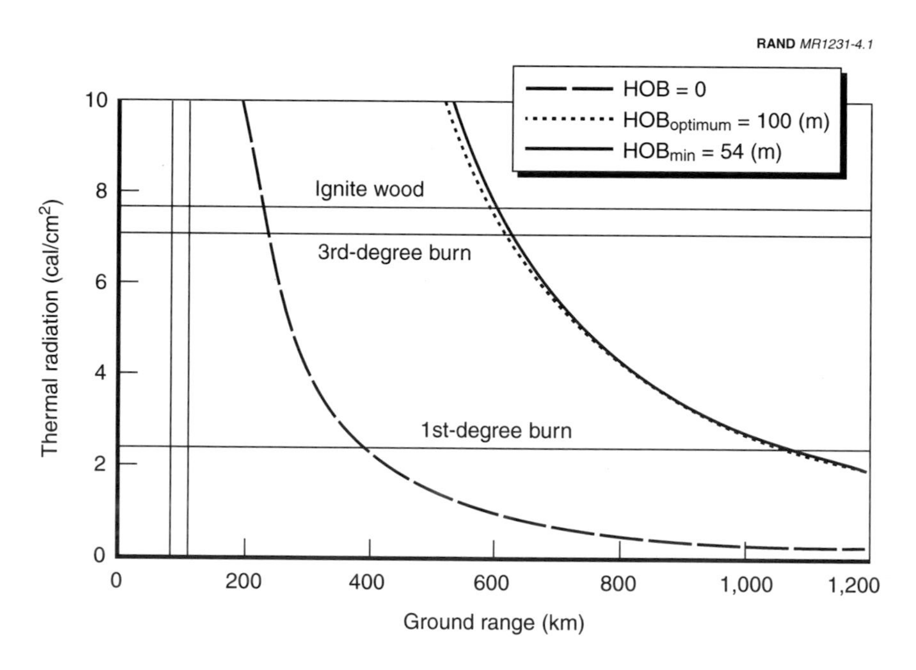

Figure 4.1—Thermal Radiation Versus Ground Range for 1-kT Weapon

RAND MR1231-4.2

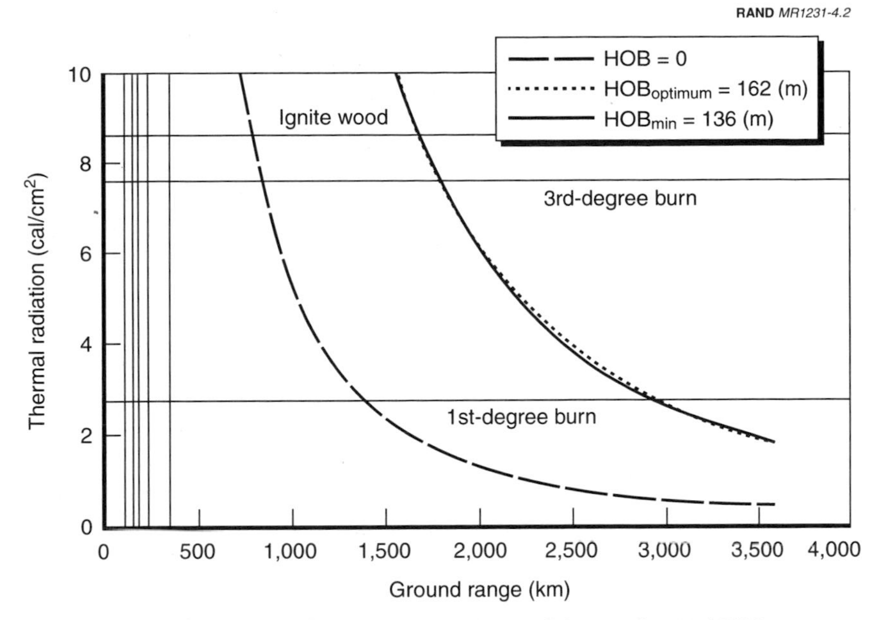

Figure 4.2—Thermal Radiation Versus Ground Range for 10-kT Weapon

RAND MR1231-4.3

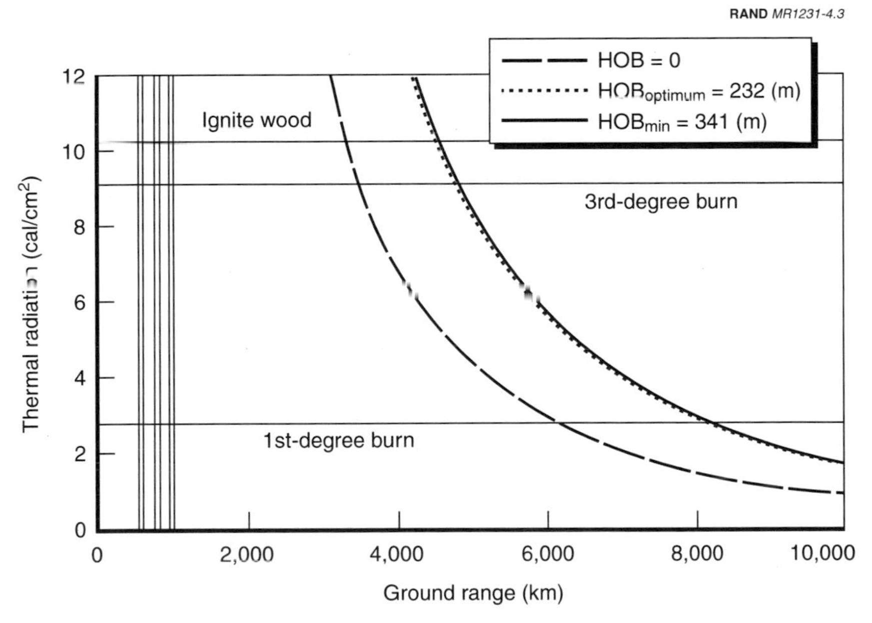

Figure 4.3—Thermal Radiation Versus Ground Range for 100-kT Weapon

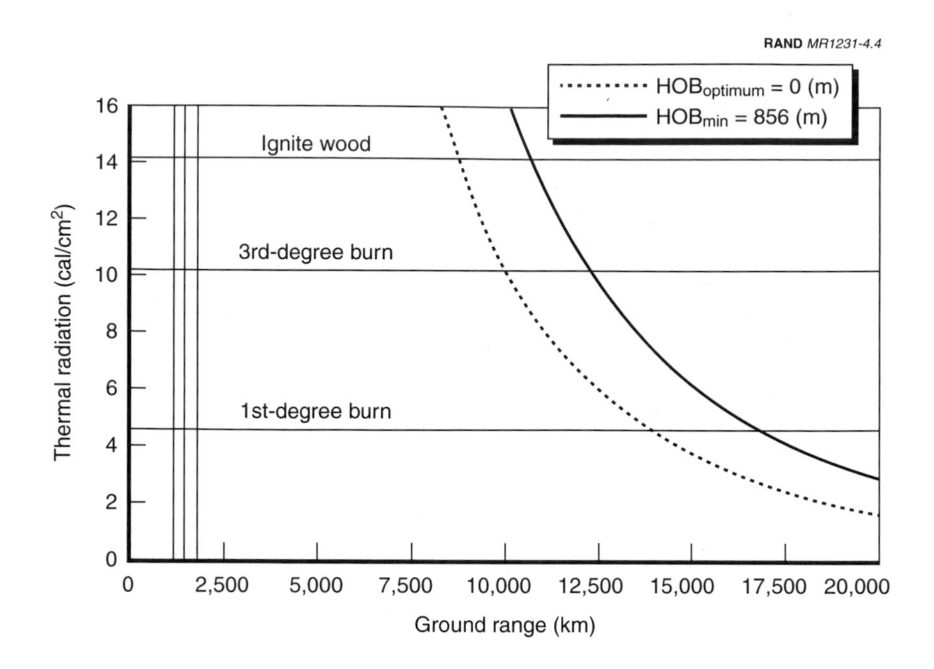

Figure 4.4—Thermal Radiation Versus Ground Range for 1000-kT Weapon

The 1-kT curve shows that exposed personnel within 1 kilometer (km) of an airburst will receive more than a first-degree burn, an undeniably serious level of collateral damage. The 1000-kT curve shows that sufficient thermal radiation is produced to ignite wood 10 km from ground zero, even for a ground burst. This level of collateral damage clearly cannot be ignored, unless the battle occurs in an unpopulated area (e.g., a desert).

To compare these two levels of collateral damage, consider the areas over which the thermal radiation from each of these weapons can ignite wood. For the 1-kT weapon, this area is $\pi\ (0.6\ \text{km})^2 \approx 1\ \text{km}^2$. But for the 1000-kT weapon, wood can be ignited over $\pi\,(10\ \text{km})^2 \approx 300\ \text{km}^2$. Of course, the seriousness of the collateral damage will depend on what is within these areas (the fires these thermal levels can produce will be far more destructive in built-up areas than in a desert).

Halting an Army: Pros and Cons of Nuclear Versus Conventional Weapons

We can now examine some of the pros and cons of nuclear weapons versus smart weapons in halting an invading army.

First and most obvious, nuclear weapon damage to an armored unit is generally far in excess of that produced by SFW or BAT—either a pro or a con for nuclear weapons depending on the objective. If the objective is to thoroughly destroy a large number of tanks and kill their crews, then nuclear weapons are far more effective than any conventional weapon, smart/brilliant or otherwise.[3] However, if the objective is simply to halt an invading armor unit, then SFW/BAT can probably do the job, and nuclear weapons may be more destructive than necessary. SFW and BAT will produce much less collateral damage or fratricide than nuclear weapons.

The platform requirements for nuclear versus smart/brilliant weapon delivery also shed light on the pros and cons of these weapons. First, whatever the platform, nuclear weapons have the advantage over *non*-smart/brilliant, conventional weapons in halting an armored invasion because of the much larger effective radius of a nuclear weapon versus the relative inaccuracy of the "dumb" conventional munitions. Hundreds of sorties with dumb conventional weapons would be required to achieve the same objective as with a given nuclear weapon. That is why tactical nuclear weapons were developed for this role during the Cold War.

However, for smart/brilliant conventional weapons, the situation is not so simple. Let us consider the case of B-2 bombers being used to halt an invading army.

The number of B-2 bombers needed to halt an invading army using conventional versus nuclear weapons was compared in previous

---

[3]On the other hand, the most effective way to use nuclear weapons against armored forces is to deliver a large enough dose of neutron radiation to incapacitate tank crews immediately. Only tanks that are relatively close to the blast are likely to be damaged severely. By contrast, precision-guided conventional weapons are designed primarily to cripple the vehicles themselves, although some are also lethal to the tank crews. Thus, there is a basic difference in the *kind* of damage that conventional and nuclear weapons inflict on armored forces. Under some conditions, conventional weapons may cause greater damage to vehicles than would nuclear weapons.

RAND research. Simulations were made of an attack on an armored division of three regimental columns in road march formation. Each column was 12 to 15 km long and separated by about 10 km. A total of 750 vehicles were in the division.

Three B-2s, each carrying 32 SFWs, destroyed 350 armored vehicles, or 47 percent of the division (note that in this study BAT was found to be more effective than SFW). This percentage of destroyed vehicles was considered sufficient to halt the division.

An armored force in a road march formation should be easier to attack than one that is deployed in an attack formation. In fact, another RAND analysis shows that in an attack against the same armored division, now deployed in an attack formation, three B-2s armed with SFWs destroyed more than 250 armored vehicles (250/750= one-third of the division).[4] Given the shock value of SFWs (the weapons arrive within a few tens of seconds), destroying this number of vehicles was still considered sufficient to halt the attack.

Elsewhere, we have calculated the number of nuclear weapons needed to accomplish this same mission, that is, the number of weapons which would be capable of stopping the division (and probably destroying it to a much more severe level than with SFW) when itis in a road march formation.

In an attack formation, there will be more dispersion between vehicles than in a road march (100 to 500 m separation versus 10 to 50 m). The vehicles are dispersed by getting off the road and traveling cross-country. However, they will still be concentrated (i.e., the entire division will still be concentrated in one area, even though the spacing between individual vehicles will increase by about a factor of ten). Therefore, given that a nuclear weapon causes damage over a much larger area than a conventional weapon (see the companion report for comparison of the area covered for a given level of damage), an armored unit dispersing into an attack formation does not pose as great a challenge to nuclear weapons as it does to conventional smart/brilliant weapons. For the same number of vehicles, only a few more warheads would be needed to effectively halt an attack than to halt a division in road march formation.

---

[4]Buchan et al. (1993).

We found that even very-low-yield nuclear weapons are as effective at destroying columns of tanks in road march (or tanks dispersed in an attack formation) as the full complement of SFW above. Also, the use of low yields could reduce the burden of fission products in comparison with relatively high-yield weapons. The difference in fission products between the two cases depends on the yield ratios being compared and the physical dispersion of the weapons and targets. The range of significant blast and thermal damage is found to be relatively small with the use of very small weapons. Collateral damage is limited to the proximity of the area attacked. For example, Figure 4.1 shows that for a 1-kT weapon, third-degree burns occur to a distance of only about 200 meters (depending on atmospheric conditions).

Two important points: First, the B-2 must be at a safe distance to avoid nuclear weapon effects, and so must either leave rapidly or remain at sufficient standoff. Second, delivery accuracy is not as important with nuclear weapons as with SFW (which is valuable because it allows the B-2 to maintain greater standoff).

To summarize the pro-nuclear weapon results for B-2s: First, a small number of B-2s armed with nuclear weapons should be capable of halting an armored division. This result favors nuclear weapons if SFW/BAT are not available in sufficient quantities, but nuclear weapons are, if SFW/BAT prove less effective than expected, or if there are not enough platforms available to deliver the number of conventional weapons needed to be effective. This could happen if:

- The United States does not buy enough advanced conventional munitions.
- The United States has an inadequate supply of suitable delivery platforms (a distinct possibility unless the United States either buys a larger bomber force or can assure itself of rapid access to local bases).
- Enemy countermeasures reduce the effectiveness of SFW/BAT.

Second, delivery accuracy is much less important for nuclear weapons than for SFW, at least for the yields considered here. The precision with which targets must be located is reduced, as is the quality of surveillance support required. The advantage for low-yield nuclear weapons is best found when combined with precision guidance achieving accuracies better than 100 meters.

The use of very-low-yield weapons against armor could potentially be quite effective. The lethal range does not scale linearly with a reduction in yield. For example, a 1-kT warhead detonated at low altitudes has a lethal weapon effect radius that is a substantial fraction of a 10-kT warhead's lethal radius. Subkiloton weapons would also have a substantial lethal radius.

The argument against nuclear weapons is basically the following: A modest force of SFWs can halt an armored division. That is good enough, even though nuclear weapons would be more effective. Moreover, using conventional weapons eliminates all of the baggage associated with nuclear weapons—e.g., release authority, nuclear training for air crews, collateral damage, diplomatic impact.

Also note that for either nuclear weapons or SFW, a time-critical target scenario requires a bomber, not a long-range missile (unless the missile can use inflight updates). For example, assuming a 40 km/hr road march, each regimental column can move 20 km within a missile flight time of 30 minutes. This distance is greater than the length of the columns themselves (12 to 15 km per regimental column). Therefore, to cover both a regimental column and the length of road over which it might move after the missile is launched would require covering (15 + 20 =) 35 km. Therefore, to cover this length of road for each of the three regimental columns would require more than twice as many weapons when we account properly for columns' movement. This result also assumes that the attackers are able to predict the position of the columns 30 minutes after the missile is launched, which requires knowing what roads the columns will be using and predicting how the vehicles will move.[5]

The speed of an armored division in the attack will be less than for a road march (about 15 km/hr versus 40 km/hr). Within a missile flight time of 30 minutes, the attacking division should move about 8 km. Therefore, a division in the attack scenario is not as time critical for targeting as a division in a road march. Also, once deployed for

---

[5]Relying on preplanned "killing zones" (i.e., specific areas in which the United States plans to engage invaders) has always been a preferred U.S. tactic.

the attack, an armored division's objective is likely to be evident. Therefore, its direction of movement should be easier to predict than a road march formation. It may be realistic to consider using a missile in this case.

These estimates (especially those of the nuclear weapon effects) are fairly rough. The nuclear weapon capabilities especially should perhaps be examined in greater detail to account for the possible impact of nuclear weapons tailored to halting invading armies (e.g., enhanced radiation, very low yield, etc.). It would also be worth examining nuclear effects (related to halting invading armies) for other damage levels than those specified by the PVH to establish in greater detail how Air Force nuclear weapons may be effective in scenarios outside their traditional roles.

STRATEGIC NUCLEAR WEAPONS IN DESTROYING HARDENED BUNKERS CONTAINING WMD

A second target class we examined was a bunker possibly containing some type of WMD: chemical, biological, or nuclear. For analytical purposes, we assume the bunker is structured as described in the PVH—that it is an arch-shaped, reinforced-concrete igloo with earth cover. We note that this type of structure is distinct from a deeply buried facility (which will be discussed later). The specified level of damage inflicted on the bunker is collapse of the arch and/or the blowing in of the end wall and door, and light-to-severe damage to the bunker's contents.

Obviously, a large enough nuclear weapon, accurately delivered, is capable of destroying such a hardened bunker. However, as in the case of halting an army, recent developments in conventional weapon technology have made it possible for a relatively small amount of air power to destroy hardened bunkers without resorting to nuclear weapons. Therefore, advanced technology again forces us to consider carefully the tradeoffs between conventional and nuclear weapon use.

To examine the problem of using nuclear weapons to destroy hardened bunkers, we will review what nuclear weapons can do against these targets (i.e., how nuclear weapon effectiveness against hardened bunkers correlates with yield, HOB, and delivery system CEP).

Next, we compare the nuclear capabilities with the effectiveness of advanced conventional weapons designed for destroying hardened bunkers (i.e., precision guided weapons [PGWs]). Similarly, we compare the advantages and disadvantages of nuclear versus conventional weapons in destroying bunkers. Finally, we will consider the possibility of developing a new hardened-bunker-destroying weapon that incorporates the positive features of both nuclear and conventional weapons.

Destroying Bunkers: Nuclear Weapons

To assess the capability of nuclear weapons to destroy hardened bunkers, we again used effectiveness values calculated from the PVH. Whereas the PVH included radiation in calculating effects of nuclear weapons against tanks, only blast effects are included in the values for damaging bunkers. We calculated values for probability of damage against the bunker design described above. To take into account possible errors in weapon delivery, we used CEPs representative of ICBMs and SLBMs. These values were chosen to facilitate calculations using the tables and figures in the PVH. (The data in the PVH did not allow efficient calculations using typical cruise missile CEPs [e.g., ~ 10 m] for the yields considered here. But certainly, at least for a single bunker, our probability of damaging the bunker to the desired extent would increase for a delivery system with such a small CEP.) By using modern guidance technology, the CEPs of future U.S. nuclear weapons can be made almost arbitrarily small. As in the scenario of halting an invading army, we compare weapon effectiveness in the two cases of a HOB selected to maximize the area affected ($HOB_{optimum}$) and a HOB high enough above ground to avoid appreciable fallout (HOB_{avoid}).

In other work, results show the probability of damage to a bunker (i.e., the probability of damaging the bunker to the degree specified above) at a given distance from desired ground zero (DGZ). Not surprisingly, the higher the yield and the smaller the CEP, the better the chance of destroying a bunker out to a distance from the DGZ. The effect of decreasing CEP can be seen in that other work as well.

Destroying Bunkers: Conventional Precision-Guided Weapons

To compare conventional weapon effectiveness in destroying bunkers, we used the results of a previous RAND analysis of hardened aircraft shelters (i.e., bunkers) attacked at three bases by precision-guided weapons during the Gulf War.

The bunkers examined were similar to (or stronger than) the generic bunker described in the PVH in terms of basic construction. All bunkers were constructed of reinforced-concrete arches with earth cover. In addition, some of the bunkers had a concrete slab covering the earth layer. The aircraft used was the F-117. The F-117 carried the GBU-27.

Mission success was defined as a mission that results in (1) one or more aircraft killed inside the bunker, (2) damage to the structure requiring repair to restore use, or (3) both (1) and (2). The mission success rate for the weapons dropped at the three bases is given in other work. A similar probability of damage from a 100 kT-warhead (detonated with HOB sufficient to avoid significant fallout) covers a large circular area.

Destroying Bunkers: Advantages and Disadvantages of Nuclear and Conventional Weapons

The principal advantage of nuclear weapons over conventional PGWs for destroying bunkers is the ease with which a single nuclear weapon can cause similar destruction over a much larger area. Covering such a large area with the desired level of damage probability permits destruction of several bunkers with a single weapon, depending on the spacing between bunkers. Therefore, targeting accuracy is nowhere near as important for nuclear weapons as it is for PGWs in destroying bunkers.

The principal disadvantage of nuclear weapons in destroying bunkers, as compared with PGWs, is the possibility of collateral damage, particularly from fallout (representative values for fallout are discussed later). Moreover, minimizing fallout does not minimize blast or thermal collateral damage.

Likewise, the principal advantage of PGWs over nuclear weapons for destroying bunkers is the possibility of obtaining extreme accuracy with little or no direct collateral damage. However, the plume from an accidental VX nerve gas release (for example) can be substantial, with potential unsafe levels extending tens of kilometers. For an assured level of destruction of a given number of bunkers, more sorties will be required with PGWs than with nuclear weapons.

Destroying Bunkers: "New" Weapons?

These results seem to indicate a possible rationale for developing a nuclear warhead small enough to be carried by a Guided Bomb Unit (GBU) similar to those used in the Gulf War. Because the warhead would be delivered with such high accuracy, it could be of small yield (< 1 kT), and yet produce a significantly higher probability of completely destroying a bunker (and its contents) than either a conventional PGW or a standard nuclear weapon. Such a small, accurately delivered nuclear warhead may also lower the probability of collateral damage. However, some local fallout will occur.

Finally, it might be possible to develop a new, unconventional nonnuclear weapon combining the best features of both: the accuracy of a PGW and a heat/thermal effect nearly that of a nuclear weapon (with no radiation/fallout) to ensure destruction of a bunker's contents. However, further analysis must determine if such a weapon's tailored nonnuclear effects could be superior to those of a nuclear weapon.

STRATEGIC NUCLEAR WEAPONS IN DESTROYING A DEEPLY BURIED COMMAND AND CONTROL FACILITY

Another Air Force application of nuclear weapons is the destruction of a deeply buried structure containing a command and control (C^2) facility.

We consider a deeply buried C^2 facility here to be characterized by tunnel facilities at depths beyond a certain number of meters. The tunnels are assumed to be constructed through rock. The nature of this rock may range from a porous and soft variety, such as limestone, to a hard type, such as granite. The estimates for hardening and depth of typical deeply buried potential adversary sites are given in the companion report, along with the estimated maximum stress survivable by a buried structure.

The configuration of a deeply buried facility is unlikely to be known accurately. The depth, length, and layout of the tunnels (i.e., the facility's "floor plan") are likely to be uncertain. Our knowledge of a deeply buried facility's configuration will be highly dependent on the quality of intelligence available.

The level of destruction visited on a deeply buried facility can vary from simple disruption of operations within the facility (to include breaking its communications link with the outside) to the collapse of the tunnel structures and the destruction of the facility's contents. The latter level of damage would provide us with a greater assurance that the facility is no longer a threat and cannot be reconstituted.

If any weapon is capable of accomplishing this level of damage, it is a nuclear weapon. However, as with the scenarios discussed so far, there are tradeoffs between using nuclear and conventional weapons in destroying (or attempting to destroy) a deeply buried C^2 facility.

To examine these tradeoffs, we review here the capabilities and limitations of nuclear weapons at destroying deeply buried facilities. We also briefly discuss what conventional weapons may be able to do against these targets, even if the conventional weapons' capabilities fall short of destroying the interior of a deeply buried facility.

Destroying Deeply Buried Facilities: Earth-Penetrating Conventional Weapons

For the highest level of damage—collapse of the tunnel structures and the destruction of the facility's contents—current earth-penetrating conventional weapons do not reach sufficient depth. Although such weapons are useful for some tasks (e.g., destroying bunkers of the type previously described), they simply cannot penetrate far enough into rock for their relatively small conventional warheads to have much of an effect against a deeply buried structure (especially one of unknown configuration). For example, tests of steel-rod penetrators reached a depth of two meters in concrete at 1200 meters per second.[6]

Destroying Deeply Buried Facilities: Nuclear Weapons

Therefore, we turn to nuclear weapons. The primary effect that enables a nuclear weapon to destroy a deeply buried facility even if it cannot penetrate the facility itself is the ground shock produced if the weapon's blast is even partially coupled to the ground.

To examine this effect, we show here peak overpressure versus depth for 1, 10, 100, and 1000 kT warhead detonations in hard rock (Figures 4.5 through 4.8). These curves are calculated from data in the PVH

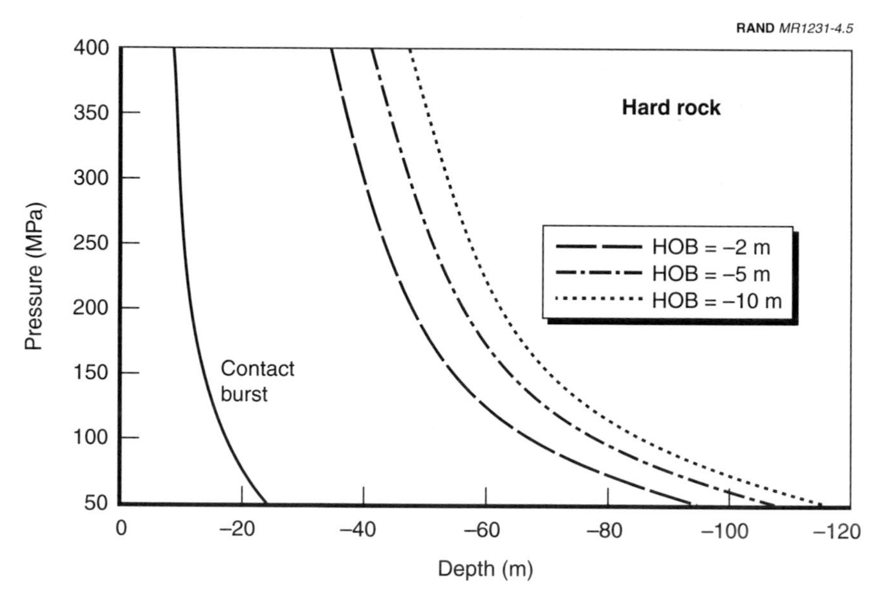

Figure 4.5—Peak Overpressure Versus Depth for 1-kT Warhead

---

[6]United States Air Force (1995), p. 26.

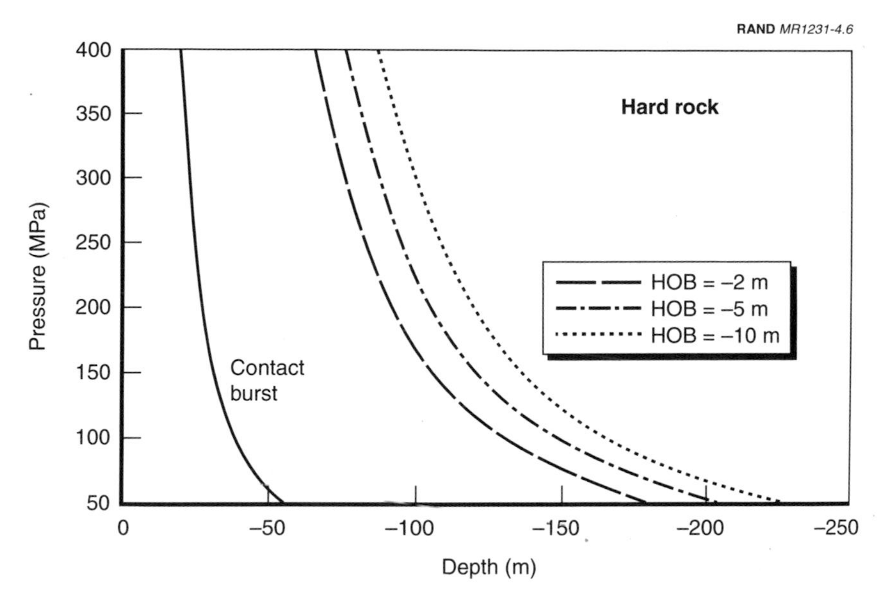

Figure 4.6—Peak Overpressure Versus Depth for 10-kT Warhead

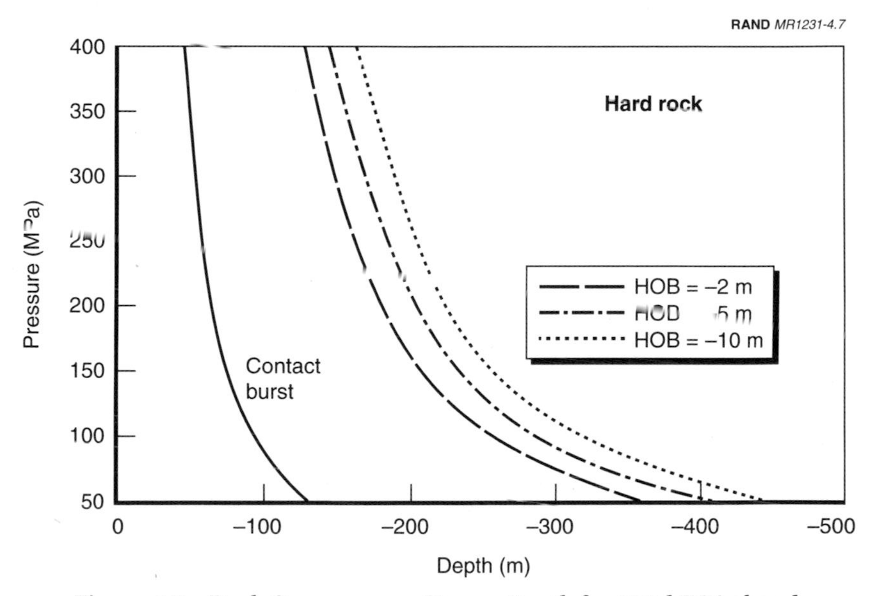

Figure 4.7—Peak Overpressure Versus Depth for 100-kT Warhead

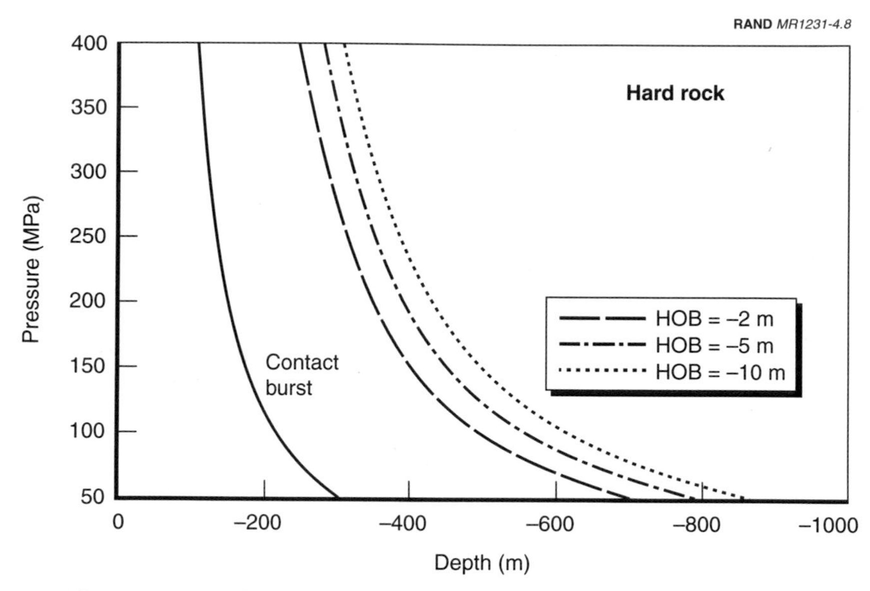

Figure 4.8—Peak Overpressure Versus Depth for 1000-kT Warhead

for ground shock. For each yield, we plot the overpressures for a contact burst and for warheads buried at 2, 5, and 10 meters. To see the effect that different rock densities have on the pressure levels at depth, we also show the curves for a 100-kT warhead exploding in porous rock (Figure 4.9).

An assumption in the PVH data is that the buried warhead is surrounded by rock of uniform composition and density, and that this rock is uniform all the way down to where the overpressures are measured. For a real target, therefore, the pressure wave may not be as strong at a given depth as shown here. For example, the warhead may detonate in soil or sand, which will attenuate some of the shock wave before it reaches the underlying rock. The same effect would hold true for a contact burst. Also, the warhead will not be perfectly embedded in the rock. That is, the material behind the warhead will be disturbed and therefore not of the same density as the rest of the rock. The net result of these factors is that, because the pressure wave is strongly affected by variations in the material through which it passes, in a real situation, pressures and/or distances may vary by a factor of 2 to 10 from those estimated here.

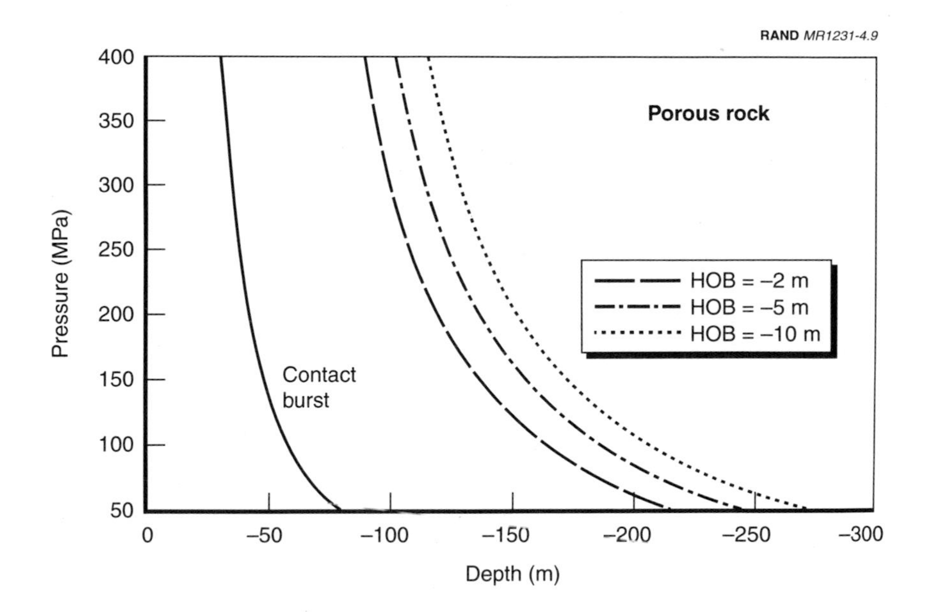

Figure 4.9—Peak Overpressure Versus Depth for 100-kT Warhead in Porous Rock

Nonetheless, we can see two important results from these curves: First, even shallow penetration matters. For example, for a 100-kT warhead, going from a contact burst to a 2-meter penetration moves the 1-kbar pressure contour about 150 meters deeper. Second, high yields are necessary for some conceivable targets. For example, assume that a facility located at a depth of 250 meters is attacked with a weapon that penetrates to a depth of 2 meters. Then, if we assume at least 100 MPa of peak overpressure is necessary for confident destruction, examination of Figures 4.5 through 4.8 shows that at least a 100-kT weapon must be used to destroy the facility with confidence. Also, if a penetrating weapon was not available, only a 1000-kT warhead (surface burst) could have any effect on the facility (and, given our assumptions, would not necessarily destroy it).

Either an ICBM or bomber could deliver the nuclear weapon. There are tradeoffs involved in selecting the most desirable platform. Either high speed at impact or a low speed with a sufficient mass density is useful for penetration. Strong materials and long length-to-diameter ratios are also factors in penetration.

We next want to examine how our less-than-complete knowledge of a deeply buried facility's configuration limits our ability to destroy it. To examine this limitation, we plot in Figures 4.10 and 4.11 peak overpressure contours at depth and range from a 100-kT warhead detonated both at the surface (a contact burst) and three different depths of burst for penetrating warheads (i.e., the weapon was detonated near the upper left-hand corner of each plot) for both porous and hard rock.

The results show the limited effect of even an embedded nuclear weapon on an underground structure of unknown configuration. An uncertainty of even a few hundred meters in the structure's configuration can preclude its destruction. Although higher-yield weapons will help improve our chances of destroying a facility, even they do not compensate for a lack of targeting intelligence. Nuclear weapons are not a perfect solution for deeply buried targets, but they are

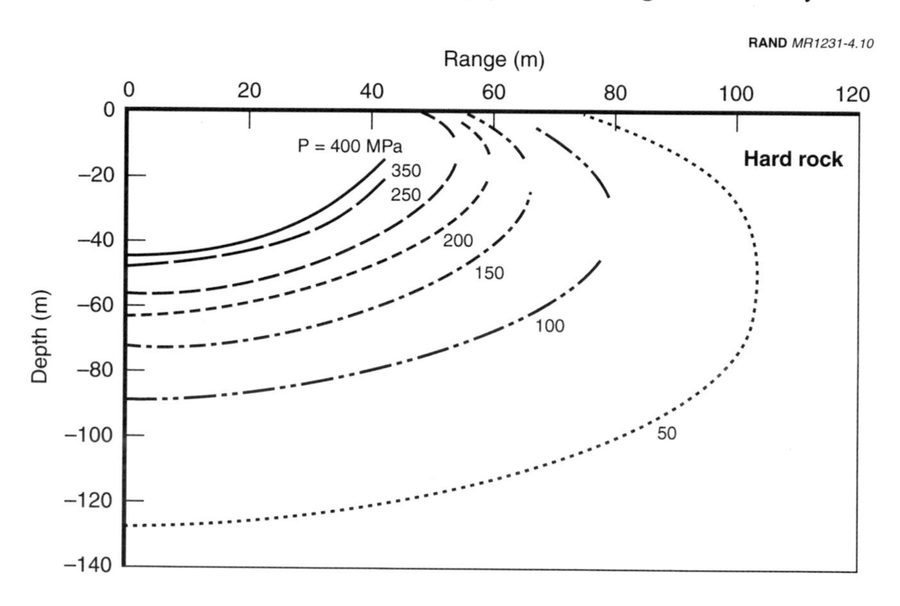

Figure 4.10—Peak Overpressure at Depth and Range for 100-kT Contact Burst

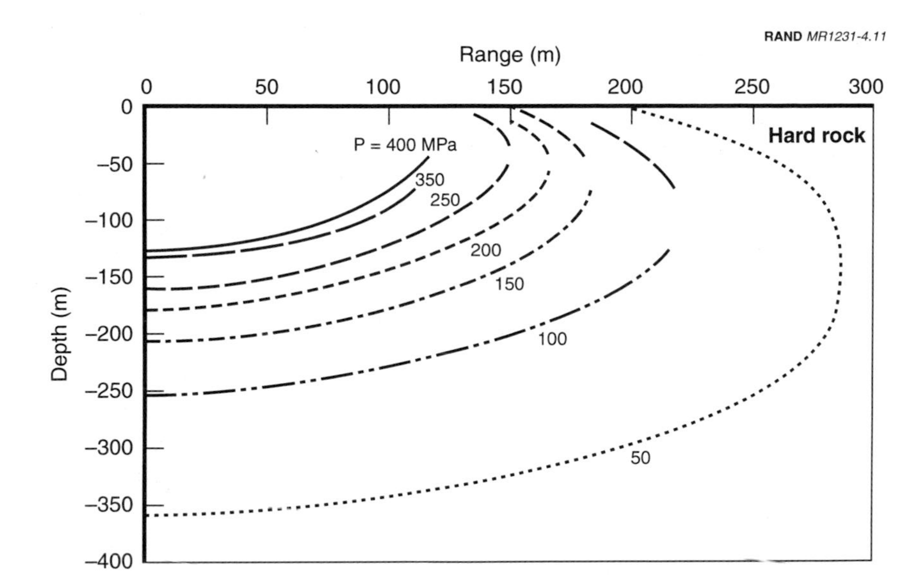

Figure 4.11—Peak Overpressure at Depth and Range for 100-kT Warhead Detonated 2 m Below Rock Surface

probably the best solution. However, there are limits to what can be done.

Destroying Deeply Buried Facilities: Collateral Damage

Another major disadvantage of using nuclear weapons to destroy deeply buried targets is that the effects of a weapon detonating a few meters beneath the surface will not be confined underground. As with any nuclear weapon exploded on or near the surface, fallout will be produced. The area covered by fallout depends on the yield and, as we found, in some cases only high-yield weapons will provide assurance of destroying certain deeply buried targets.

To show how severe such fallout could be, we plot in Figures 4.12 and 4.13 the radiation dose (in rads) as a function of downwind range under mild winds for a few yields (in Figure 4.13, the vertical axis is expanded). These curves show the dose received after 72 hours of exposure to the fallout from a surface burst. We also show the phys-

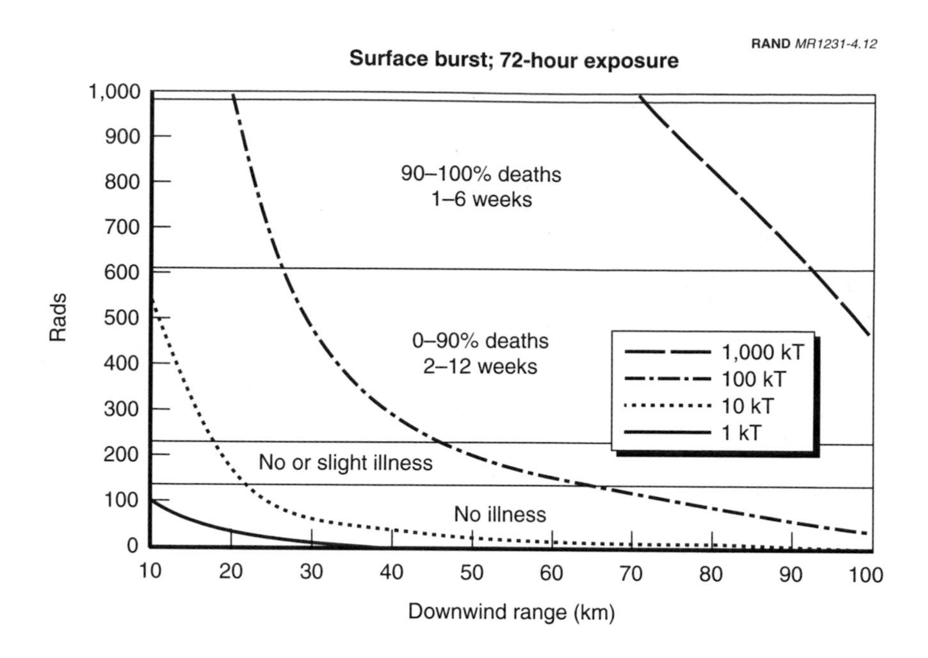

Figure 4.12—Fallout Dose Versus Downwind Range

iological effects of these dose levels. The severe fallout effect from a 1000-kT weapon (nearly 90 percent of persons within 100 km from ground zero would die) would seem to preclude using such a yield to destroy a deeply buried target.

Destroying Deeply Buried Facilities: Nuclear Versus Conventional Weapons for Functional Kills

Given the fallout effects associated with using nuclear weapons to destroy deeply buried facilities, conventional weapons may be the best bet for a functional kill of a deeply buried C^2 facility. A functional kill may suffice to eliminate the facility as a threat or at least provide temporary disruption of its activities. A functional kill can be accomplished by destroying above-ground features of the C^2 facility such as antennas, landlines, vents, etc. Although such features are, by themselves, easy to destroy, it is difficult to ensure that they have

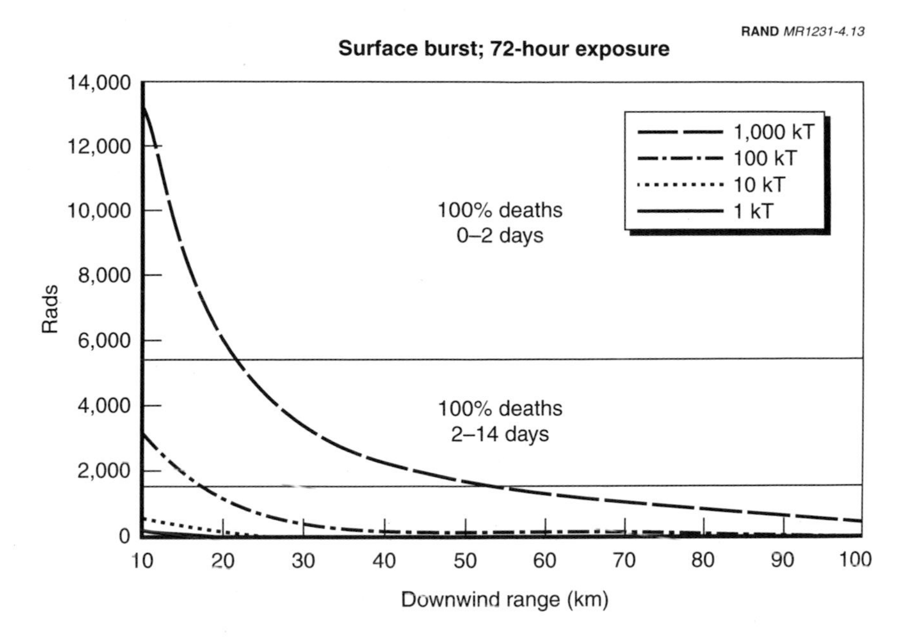

Figure 4.13—Fallout Dose Versus Downwind Range (Vertical Axis Expanded to Show Fallout from Higher-Yield Weapons)

all been eliminated and that the facility is not continuing to function through the use of backup resources. Also, while conventional PGWs are well-suited to attacking such targets, they will require very accurate target location.

Nuclear weapons would not require as accurate targeting as would PGWs to destroy a facility's above-ground features. A few lower yield (<10 kT) nuclear weapons, detonated at sufficient HOB to minimize fallout, may provide a functional kill of a deeply buried C^2 facility.

ROLE OF NUCLEAR WEAPONS IN DEFENSE AGAINST BALLISTIC MISSILES

The application of nuclear weapons to destroy offensive missiles in flight is promising from technical considerations and may serve as a special option among missile defenses. There are several advantages:

- Relax interceptor accuracy requirements
- Destroy biological warheads
- Rapid deployment
- Reduce kill-assessment uncertainties.

Two scenarios in which nuclear weapons might be used are (1) a theater situation where an adversary employs or threatens to use ballistic missiles armed with WMD, and (2) the threat of a limited ICBM attack on the continental United States (CONUS) by an emerging power. Kinetic-energy weapons must hit or come very close to their targets to kill them. The development of the sensors and guidance needed to achieve this might at first seem to reduce or eliminate the need for defensive nuclear warheads, but nuclear warheads can increase confidence in the effectiveness of the system. The development of accurate interceptors also makes possible consideration of the use of very small nuclear warheads, which can greatly reduce unwanted side effects. Collateral damage, political considerations, and tactical disadvantages (e.g., self-induced blackout) have been the main drawbacks to nuclear weapons in a defensive antimissile role. Combining nuclear warheads with modern technologies may alleviate some of the drawbacks.

Missile Defense: Nuclear Weapon Performance

Of the three main types of weapons of mass destruction—nuclear, chemical, and biological—chemical warheads are probably the most vulnerable to conventional missile defense but still pose problems for kill assessment. How can we know that the target was sufficiently disrupted and spilled any chemical contents before reaching its destination? How can we ensure that a hit-to-kill vehicle destroyed the toxic contents of a biological WMD? What if the adversary uses nuclear warheads and the defense's probability of kill must be extremely high?

All of the early ballistic missile defense schemes relied on nuclear warheads. The discussion below on the nuclear threat highlights the issues. Growing recognition of the biological threat and the potential to counter it with nuclear warheads provides a strong incentive to consider the application of nuclear weapons in this role.

Nuclear Threat

The neutrons released during the detonation of a nuclear warhead can induce a sufficient number of fission reactions to melt fissionable material in another nuclear warhead at close distance. For a given yield, an enhanced-radiation warhead would be more effective than a fission warhead, but even a more typical modern nuclear warhead could achieve a sure kill in proximity to an adversary's nuclear warhead. A precise calculation of the range versus yield requires a model of design details of the warhead on the incoming missile.

However, a representative calculation can be based on first principles without detailed knowledge of nuclear weapons design. The critical mass of fissile materials is well-known to require an assembled thickness of some number of centimeters. Disregarding the particular geometry of a weapon, if its fissionable mass is significantly deformed by heating, then a successful nuclear explosion can be denied to the adversary.

There will be little or no atmosphere to absorb the neutrons in a missile defense scenario. Geometric dilution of neutrons is the most important factor for the neutron kill performance of nuclear warheads in missile defense, not atmospheric absorption.

The generic fluence on the target missile at a given distance in meters as a function of yield (Y) in kilotons is approximately:

$$F(\text{neutrons}/\text{m}^2) = (1.5 \times 10^{22}) Y(\text{kT}) / R(\text{m})^2$$

The fissionable material must be heated to its melting point and then further heated to undergo a phase transition. Uranium melts when the energy density is about 200 joules per gram.[7] The kill range for a neutron mechanism is proportional to the square root of yield:

$$R \sim Y^{0.5}$$

Although larger weapons have greater absolute kill range, a desire to minimize collateral damage may favor smaller weapons, because the

[7]Wiegel (1992), Volume 19, p. 86.

loss in kill range is not linear with yield, and small weapons retain appreciable kill range relative to larger weapons.

The ratio of the mass of a uranium nucleus to the combined mass of fusion isotopes is high, so the number of neutrons released per gram by fusion is much greater than the fission ratio of neutrons per gram. However, fusion also releases higher energy yield per gram than fission. If a defensive nuclear weapon derived a portion of its energy from fusion, the kill range would be significantly enhanced over a pure fission weapon for a given yield.

Countermeasures must be considered too. Neutron absorbers could be added for hardening to prevent neutron kill. The effectiveness of such countermeasures is likely to vary depending on the sophistication of the country posing the threat.

If the ability of a BMD interceptor to achieve small miss distances were uncertain, there would be a strong incentive to use a radiation-enhanced warhead, or alternatively, a very large warhead. A large warhead can kill by other means than neutrons and is probably not required. A small warhead, such as the primary of a larger warhead, is a good candidate because of modern developments in guidance. Even if many of the interceptions were near misses rather than hits, the ability to closely approach the target missile would give a kill for a low-yield warhead. A Minuteman missile modified for defensive use and using command guidance is a potential quick route to a defense.[8]

If interception is to occur within a threat cloud, it will be difficult to identify the reentry vehicle.

Biological Threat

A biological warhead can be effective even if the reentry vehicle is disrupted at high altitude. Very small quantities of biological agent can lead to incapacitation.

---

[8]A nonnuclear intercept was most recently demonstrated on October 2, 1999, in a test for development of missile defense.

The intense heat of a nuclear explosion might sterilize biological agents, but it is not a sustained heat. Close proximity to the incoming missile might be required. A radiation kill mechanism might be necessary to ensure destruction of the agent.

Megarad dose levels are typically used in food sterilization. Hardy agents such as anthrax spores may require 4 megarads, or 40 joules, absorbed per gram of biological material.[9] This figure will be assumed as a conservative bound on the dosage needed to destroy wide classes of biological agents. The process of food sterilization uses gamma rays from natural sources or hard x-rays from accelerators. The photon energy from a nuclear explosion is typically much smaller (e.g., slightly over 1 keV) and can be significantly attenuated before reaching the biological agent.

Neutrons are penetrating and damaging to tissue. The effects are strongly dependent on the neutron energy spectrum. Fast neutrons in excess of 1 MeV produce a dose of approximately 1 rad if the fluence is 3.2×10^8 neutrons per square centimeter. Neutrons near 0.1 MeV require about five times as much fluence to produce the same dose.

The neutron fluence derived for the nuclear threat can be used to find the kill range against a biological threat. The lethal fluence in the absence of neutron moderation is approximately 1.3×10^{15} neutrons per square centimeter. This gives the kill range as a function of yield:

$$R(m) = 37\ (Y(kT))^{0.5}$$

The kill radius is much larger for a weapon that derives a large portion of its energy from fusion. In addition to the greater neutron fluence per kiloton, the higher-energy neutrons are slightly more damaging.

More susceptible organisms might be substantially destroyed at ranges that are two or three times greater than shown in the curve of Figure 4.14. Consequently, even a low-yield 10-kT weapon might kill

---

[9] Turnbull et al. (1998).

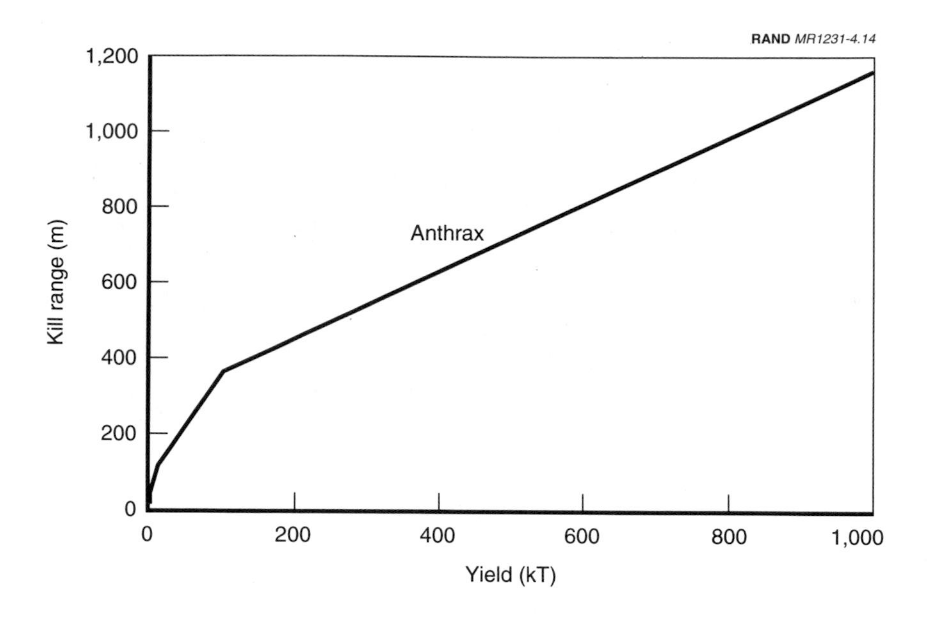

Figure 4.14—Neutron Kill Range Against a Ballistic Missile with a Biological Warhead

many types of biological agents at several hundred meters in an intercept.

A large 100-kT yield has a kill range for anthrax in excess of 350 meters. A 1-kT yield produces a kill distance of tens of meters. Low-yield weapon variants could be considered as candidates for defense against an adversary in a theater threatening the use of WMD. At high intercept altitudes, there would be no significant effects on the ground from heat or blast. Fallout would be minimal.

Missile Defense: Collateral Damage

Ground effects from blast are negligible because intercepts typically occur at tens to hundreds of kilometers. Thermal damage could result if a high-yield warhead were used at low altitudes, so this should be avoided. A 1-MT explosion at an altitude of 15 km under high visibility can produce first-degree burns at a horizontal ground distance in excess of 25 km. By contrast, a 10-kT weapon must explode at heights below 3 km to produce first-degree burns. Low-yield weapons do not produce significant collateral damage from blast or thermal effects.

Fallout effects would be limited in missile intercepts. The amount of radioactive debris produced scales with the yield. A large thermonuclear weapon does not have a linear increase because a portion of the yield is produced by fusion, but a 1-MT weapon will produce more radioactive debris than a 10-kT weapon, so the use of low-yield weapons reduces the burden on the environment.

An important source of damage might be the electromagnetic pulse from large thermonuclear explosions at high altitude. Communications might be interrupted. Scintillation effects on radar are insignificant at X-band. The effect on satellites in low earth orbit (LEO) is an important issue.

The high-altitude detonation of a defensive nuclear weapon can damage satellites in LEO or disrupt communications and electronics on the ground. Unhardened satellite electronics are susceptible to x-rays. The fraction of the yield that is released as x-rays is 70 to 80 percent for a typical nuclear warhead.[10] The x-ray fluence is

$$F(J/m^2) = 3 \times 10^{11}\, Y(kT)/R(m)^2$$

Military satellites may be hardened against x-rays, but future operations could also use or rely on civil and commercial satellites that include relatively unhardened electronics. Commercial satellites generally can be hardened against a fluence of a few tenths of a calorie per square centimeter at a small percentage penalty in dry weight and cost.[11] Additional hardening is possible at the expense of increasing the percentage of the satellite dry weight participating in hardening. Figure 4.15 presents the kill range for hardening between one and five joules per square centimeter.

---

[10] Glasstone and Dolan, p. 24.

[11] Estimate by Paul Nordon (1996), p. 221.

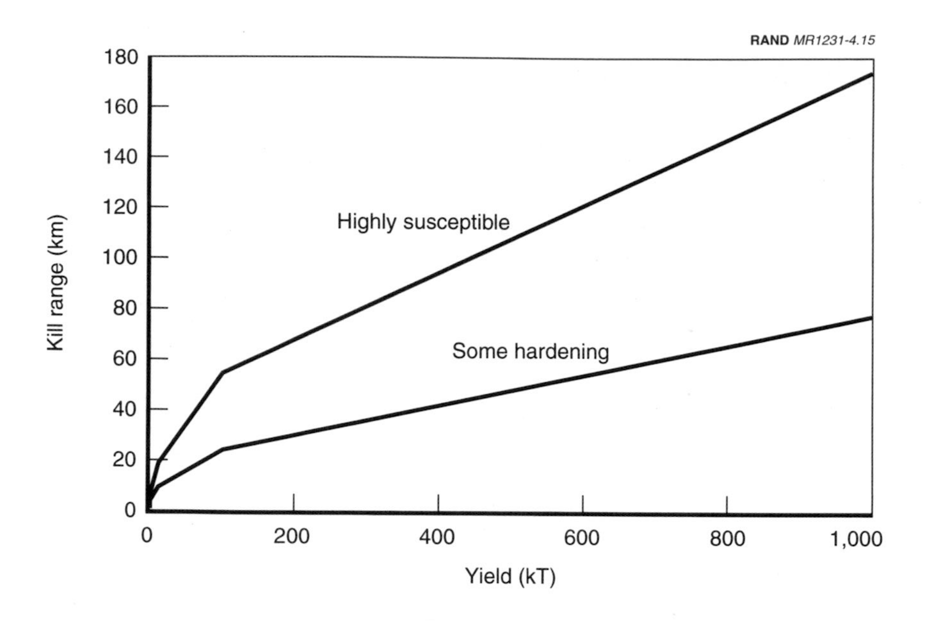

Figure 4.15—Parametric X-Ray Kill Range Against Commercial Satellites

Commercial satellites may be at risk to x-ray damage from megaton-class explosions at hundreds of kilometers in extreme cases.[12] It is thus possible that high-altitude intercepts of long-range ballistic missiles could endanger LEO spacecraft. This is not the case for low-yield explosions of a few kilotons. Only satellites within 20 km of the intercept are at risk, and if, as is likely, the intercept occurs below 200 km (the approximate cutoff for satellites to stay on orbit without decay), then no LEO satellites are at risk from the x-rays.

Neutrons can interact directly and also produce gamma rays in electronics that cause significant secondary damage. The damage to

---

[12]It is possible that a commercial satellite might be susceptible to lower thresholds than assumed above. If a threshold value for electronic components of less than 10^{-3} to 10^{-4} calories/cm^2 is assumed, then the corresponding range for possible disruptions of a satellite may be 10 to 100 times greater than shown for the "highly susceptible" curve in Figure 4.15. The practical result is that a conservative approach would consider any untested satellite in LEO within a line-of-sight of the detonation to be at some risk even from a low-yield explosion.

satellites caused by neutrons is generally less important. Transistor-based electronics are susceptible at neutron fluences from 10^{11} to 10^{15} neutrons per square centimeter.[13] The kill range for a 100-kT explosion is thus typically about 4 km. The most sensitive case involving a 1-MT warhead and the lower fluence bound gives a range of 130 km. This is less likely to be found in future satellites with good hardening practices.

Fission debris emits electrons that can be trapped in belts by the earth's magnetic field and can shorten the lifetime of satellites in a way similar to the natural radiation belts. It is desirable to minimize the total fission yield injected. A single 1-MT warhead would be equivalent to dozens of low-yield warheads. Again, weapons of at most a few tens of kilotons appear to be most suitable.

Missile Defense: Nuclear Delivery Options

High yields may appear acceptable at very high altitudes, but electromagnetic pulse may preclude this. Low yields are the best choice to limit collateral damage.

Patriot-class systems could employ nuclear warheads for added confidence.[14] However, even a relatively high-altitude endoatmospheric intercept risks local collateral damage. The amount depends on minimum intercept altitude and yield. Collateral damage from a true terminal system (e.g., Sprint) could be much worse, as shown by the studies that examined Sprint deployment at the time of the original safeguard decision.

The use of a Minuteman-based system may be the quickest route to deploying a nuclear ballistic missile defense of CONUS, but it too has disadvantages. It is desirable to restrict the altitude to protect LEO constellations, because the timing and positioning of the offense with respect to satellite ephemeris is not known in advance. Sites to be defended far from the ICBM fields require exoatmospheric intercepts. Intercept trajectories may also require detonations near

---

[13]Glasstone and Dolan (1977).

[14]SAM-D, the system that evolved into Patriot, originally included a nuclear warhead option.

populated regions. There may be situations in which some risk to satellites must be accepted. A low-yield detonation may relax this concern, although introducing fission debris as sources of high-density electron belts that cause degradation to satellites could still be an issue.

A Theater High-Altitude Area Defense (THAAD)-like system designed with good guidance to ensure sufficient proximity for kill against high-speed missiles/reentry vehicles (RVs) might be a more contemporary option. Like Patriot, it could be transportable to theaters as required.

COMPARISON: THE FOUR SCENARIO CLASSES

All four scenario classes present cases in which nuclear weapons might be used operationally. However, nuclear weapons are not uniquely suitable in all cases.

Bomber-delivered nuclear weapons could be effective at halting invading armies if all the practical operational problems were solved. A small number of B-2s could perform the mission. In the past, large numbers of sorties with conventional weapons have been required. In the future, the use of advanced conventional weapons—SFW and BAT—makes the conventional option competitive with a nuclear strike against a sudden invasion. Much smaller (1 to 10 kT), more numerous nuclear weapons than the United States currently has would be most appropriate for this sort of application. Of course, if the stakes were high enough to consider nuclear use, the United States might be willing to "make do" with strategic-sized warheads and live with the consequences.

Either conventional or nuclear weapons could be used to destroy bunkers. Nuclear weapons raise the level of confidence in doing so. The desire to limit collateral damage suggests that if a target were sufficiently important to warrant a nuclear strike, the application of a very low yield is preferable.

Only the challenging case of destroying deeply buried facilities appears to demand the application of nuclear weapons—the high accuracy and yield of ICBMs or a high-yield, accurate penetrating bomb. Large nuclear yields are generally most effective for this particular application, especially if groundshock is the only kill mechanism. Conventional weapons cannot currently penetrate through hardrock to a sufficient depth to destroy a deeply buried facility and at present could only prove effective only by attempting functional kills, such as attacking vents or communications ports of entry. That could be good enough. It is not clear how effective even nuclear weapons would be in destroying very deep facilities without good knowledge of the facility's configuration. Success depends on gathering intelligence about the internal configuration of the facility, which is generally extraordinarily difficult.

Nuclear weapons could be considered as an option for missile defense. The success of nonnuclear antimissile kill mechanisms in general could eliminate the need for nuclear warheads. However, test results to date have been breath-conserving at best. Nuclear warheads could reduce the inherent uncertainties in BMD effectiveness in cases where the stakes are high. On the other hand, the lethal radii for even nuclear warheads used in BMD applications are typically relatively modest, which means that the BMD system still has to work well enough to get the defensive warheads close to their targets.

If nuclear warheads were employed for BMD, the best option appears to be low-yield nuclear warheads that intercept incoming missiles either in the high endoatmosphere or just barely into the exoatmospheric region below the orbits of LEO satellites. This suggests a THAAD-like system capability. In a scenario to protect CONUS, the interceptors must be capable against high-speed RVs. Hit-to-kill is not a requirement, so accuracy demands are relaxed relative to the nonnuclear defensive system.

Chapter Five

IMPLICATIONS FOR FUTURE U.S. NUCLEAR STRATEGY

We have developed the general point that the United States has a variety of choices in selecting a future nuclear strategy. Choosing among the options will depend to a significant degree on technical issues such as the selected weapon comparisons in Chapter Four. In this study, we were enjoined from examining force effectiveness and operational issues in detail. However, those kinds of considerations can have a major impact on selecting a grand strategy. In this chapter, we examine some of the major strategic policy options available to the United States and some key issues that will affect the choices.

A SPECTRUM OF NUCLEAR STRATEGIC OPTIONS

The choices available to the United States run the gamut from renouncing nuclear weapons entirely to much more aggressive nuclear strategies than the United States has entertained in the recent past. Below are five generic approaches that cover the spectrum of possibilities. Options also exist to "mix and match" elements of several into combination strategies:

- Abolition of U.S. nuclear weapons, with or without formal arms control
- Aggressive reductions and "dealerting"
- "Business as usual, only smaller"

- More aggressive nuclear posture
- Nuclear emphasis.

Abolition

Doing away with U.S. nuclear weapons entirely logically must be a part of any complete set of strategic nuclear options, in spite of the fact that the current U.S. administration has made it clear that nuclear weapons remain an important part of its national security strategy.[1] Even most advocates of deep reductions stop short of calling for abolition of nuclear weapons, at least for the foreseeable future.[2] Still, there is a case to be made for abolition, and strictly speaking, the NPT commits the United States and other nuclear-armed signatories to the treaty eventually to divest themselves of their nuclear weapons.

If the United States were to choose to divest itself of its nuclear weapons, it would presumably be for some combination of the following reasons:[3]

- Lack of a military/political threat to the United States serious enough to require a threat of nuclear retaliation to deter.
- Alternatives to nuclear weapons adequate to solve any military problem.
- Conclusion that the danger, trouble, expense, and political baggage associated with maintaining nuclear weapons exceed whatever residual value they might have.
- Conclusion that nuclear weapons are not "usable" politically or militarily and that "withering away" of U.S. nuclear forces is unavoidable.
- Political judgment that giving up its nuclear weapons would do more to restrain nuclear proliferation than maintaining a dominant nuclear capability.

---

[1]Warner (1999).

[2]See Feiveson (1999) and Turner (1997).

[3]Based on Buchan (forthcoming).

The first two points are key: lack of a compelling need for nuclear weapons and the availability of adequate alternatives. The dramatic improvements in the accuracy and lethality of conventional weapons clearly make them attractive alternatives to nuclear weapons for many applications. However, they bear the burden of proof to demonstrate that they can really achieve these capabilities in practice at an affordable cost. In particular, improved information collection and processing technologies "enable" most of these weapons concepts, and not all of the requisite capabilities are in hand.

A key issue involving abolition of nuclear weapons is whether the United States would eliminate its nuclear weapons unilaterally or as part of a more sweeping agreement among several—or all—nations to abolish nuclear weapons generally. If there were actually a way to abolish nuclear weapons broadly (leaving aside for the moment the inevitable questions about verification, hidden weapons, and nations' relative ability to regenerate nuclear capability), the United States would clearly be the major beneficiary. Not only does the United States have the most to lose from a nuclear war, but also its economic and conventional military power would leave it in a position of strength in a world without nuclear weapons. The problem, of course, is that the political processes necessary to produce such an agreement boggle the mind. The bar would have to be set so high that the question becomes almost academic. If the international political climate became that benign, nobody would need nuclear weapons anyway.

A more interesting case is the one where the United States considers the possibility of unilaterally eliminating its nuclear weapons. Aside from the fact that the current political climate in the United States would not allow that drastic a step, unilateral U.S. action is more interesting to consider because that is probably the only way such a result could come about. It simply requires a U.S. decision instead of the epic multilateral negotiating process and eventual implementation and verification machinery of a formal agreement.

Finally, there is the possibility that the United States may not be able to retain its nuclear infrastructure and nuclear weapon design and operational expertise indefinitely no matter what policymakers would prefer.[4] As we and others have observed, this is a major issue. If that proves to be true, the decision for the United States will not be whether to eliminate its nuclear weapons. Rather, the only question will be when and how. We will return to this point later in this chapter.

Aggressive Reductions and "Dealerting"

A less aggressive but sill quite radical set of proposals has been put forward to reduce the size of U.S. nuclear forces far beyond what even START III currently envisions (e.g., to a level of a few hundred) and lower their alert levels dramatically so that nuclear weapons could not be used quickly. The possibility of going to smaller force levels has been on the table, at least implicitly, since the beginning of the nuclear age. Admiral Arleigh Burke made an articulate argument in the 1950s, when the key decisions that would shape U.S. strategic force structure for the next fifty years were being made, that the United States ought to limit its nuclear force to a modest number of missiles deployed on submarines. Such a "finite deterrent" force, he argued, would provide the United States with the capability to respond to a nuclear attack by launching a retaliatory attack on an enemy's cities, and that was all that U.S. nuclear forces needed to be able to do.

Burke lost that battle, but the argument has been around in one form or another ever since. Recently, a prominent Russian defense expert said it was an option that Russia might pursue in the future, structuring its nuclear forces "on the French model."[5] The United States could move in that direction as well, regardless of any formal arms control arrangements. Several contemporary U.S. authors have made essentially that argument.[6] The contemporary version is that in the current world no reasonable use of nuclear weapons would require a large number of warheads. Thus, a force of a few hundred weapons should be adequate to handle any application that could arise for the United States in the foreseeable future.

---

[4]Defense Science Board (1998); and Buchan (1994), p. 77.

[5]Based on conversations in Moscow between RAND analysts and senior Russian officials.

[6]See Bundy et al. (1993), Blair (1995), and Feiveson (1999), among others.

The dealerting proposal is more recent, although its origins go back almost 20 years.[7] It seeks to solve a different problem: accidental war. The premise is that in the contemporary world in particular, the greatest nuclear danger comes not from a surprise attack, but from an accident or mistake on someone's part. Of particular current concern are the potential vulnerabilities of the Russian nuclear forces and the deterioration of their tactical warning systems. The primary concern is that, the end of the Cold War notwithstanding, Russia still worries about the vulnerability of its strategic forces to a preemptive attack. Unable to afford to keep its SSBNs at sea and perhaps even its mobile ICBMs out of garrison, Russia might opt—as suggested earlier—to rely on being able to launch its vulnerable missiles on receipt of tactical warning of an attack in progress. That in itself is bad enough, the theory goes, especially because of the deteriorating state of Russian tactical warning systems. It increases the risk of an error, such as the incident with the U.S. sounding rocket launched from Norway,[8] that could lead to an accidental nuclear war. Even worse, lacking faith in their tactical warning systems, the Russians might feel desperate enough to launch a preemptive attack if they suspected they were about to be attacked.

Dealerting is intended to solve those problems by making it more difficult for the United States to launch a nuclear attack quickly.[9] Details of individual proposals vary. Most include separating warheads from delivery vehicles or key components from missiles so that time would be required to prepare to launch a nuclear attack.[10] The objectives would be to

- avoid a premature launch of U.S. nuclear forces prompted by a false alarm or other mistake
- allow time to think about the wisdom and nature of a nuclear response

---

[7]Blair (1985), pp. 288–295.

[8]For descriptions of this incident, see Blair (1995), p. 51; and Blair et al. (1997), among others.

[9]See, for example, Turner (1997), Feiveson (1999), and Blair (1995).

[10]Blair et al. (1997).

- reassure the Russians and other nuclear powers that the United States does not intend—and cannot execute—a nuclear attack "out of the blue." The presumption is that the Russians would reciprocate by reducing the "hair-trigger" response options of their own nuclear forces.

Normally, the U.S. nuclear weapons bureaucracy would dismiss such ideas out of hand. This time, however, the proposal was harder to ignore, particularly when it was put forward recently in a paper coauthored by prominent strategic analyst Bruce Blair and former Senator Sam Nunn. Dealerting became topical enough to be rejected explicitly by the Department of Defense in then–Assistant Secretary Warner's testimony to Congress in 1999.[11]

The dealerting proposals raise several fundamental strategic issues:

- Is the premise valid that accidental war is now a greater danger than a surprise attack?
- How likely would the Russians be to respond to U.S. dealerting initiatives by reducing the readiness of their own vulnerable ballistic missiles?
- Will delaying a response weaken the threat of a nuclear deterrent based on retaliation?
- What effect would it have on other potential uses of U.S. nuclear forces?
- Are specific dealerting proposals practical, and do they cause more problems than they solve?

The first two points are critical to the case for dealerting and amount to a judgment call. War resulting from accidents, misinterpretation of events, or unauthorized use of nuclear weapons has been a matter of intense concern since the earliest days of nuclear weapons and has received continuing attention from the very beginning. There has always been a tension between decreasing U.S. vulnerability to surprise attacks and increasing the risks of accidental war, and the balance has shifted over the years to reflect altered perceptions of the

[11]Warner (1999).

threat to the United States and the costs and benefits of the remedies available at the time. For example, for a while, the United States kept part of its bomber force on airborne alert to reduce its vulnerability to a surprise attack. It ended that practice, however, deciding that there were adequate solutions to the potential bomber vulnerability problem that were less dangerous and expensive than airborne alert. Similarly, for years the United States paid relatively little attention to some of the vulnerabilities of its nuclear command structure, probably because all of the potential solutions were unattractive. It was easier to redefine the problem and discount the potential threat than to take it seriously, spend a substantial amount of money, and still not find an entirely adequate solution. Thus, even during the most intense periods of the Cold War, the United States recognized the need for—and made—trades among the vulnerability of its forces to a first strike and the risks of unintended nuclear war.

As the Cold War was ending, the United States took its bomber force off a day-to-day alert as a part of a series of reciprocal initiatives between Russia and the United States to reduce the size and alert levels of their nuclear forces. Presumably, U.S. leaders understood that they were increasing the vulnerability of U.S. strategic forces to a surprise attack when they took these actions. They simply made the *political judgment that the risks of a surprise attack had decreased sufficiently that the costs and stress of maintaining bombers on strip alert was no longer justified.* Moreover, the climate at the time encouraged both sides to take initiatives in the nuclear arena. Thus, there is precedent for reducing the alert levels of nuclear forces and expecting some kind of reciprocity in a relatively benign political environment. The real issues are how far to go and under what conditions.

Russian reactions are crucial to the dealerting argument. The United States might argue—and has in fact[12]—that it has adequately dealt with its own command and control problems to minimize risks of accidental or unauthorized nuclear launches. If so, the only rationale for further U.S. dealerting would be to persuade the Russians that their strategic forces were not in enough danger from the United States to justify such dangerous strategies as launch-on-warning or

[12]Warner (1999).

preemptive attack. If the Russians are truly worried about U.S. motives, they might refuse to respond to anything other than transparent dealerting measures. (They also would have political incentives to make the price to the United States as high as possible.) There is no way to be sure that Russia would respond to U.S. dealerting initiatives unless the political climate improved to the point where such measures were almost irrelevant.

Even if the Russians accepted some sort of reciprocal dealerting arrangements, confident verification of some kinds of dealerting actions would be extraordinarily difficult. Absent that confidence, the United States could never be certain that dealerting had solved the Russian side of the hair-trigger problem. Indeed, the uncertainties and sensitivities could exacerbate rather than ease the problems that dealerting are intended to solve. This is the classical arms control dilemma of trying to do more with agreements than the political traffic will bear.

The effect of a delay in a U.S. nuclear response is clearer. *There is nothing about a strategy of deterrence based on nuclear retaliation that requires a prompt response.* There never has been. The only rationale for a quick response during the Cold War was the fragility of U.S. forces and command and control systems. The choice might have been between a quick response and no response at all. Nothing about the target base ever required a quick response,[13] which is even more true in the current world. Indeed, as we have discussed in other sections of this report and elsewhere, a prerequisite for credible contemporary deterrence enforced by a threat of nuclear retaliation is certainty about what happened and who is to blame. That puts a premium on being able to delay a response. Thus, *a dealerted U.S. nuclear force, assuming it can be made survivable, should still be capable of enforcing a strategy of deterrence based on a threat of nuclear retaliation.*

Dealerting would eliminate some nuclear strategic options. It would, of course, preclude a strategy of launching a preemptive nuclear attack "out of the blue." That, after all, is the purpose of dealerting. It would *not* preclude first use of nuclear weapons in a slow-developing

---

[13]"Prompt second-strike counterforce" never made any sense when U.S. and Soviet arsenals were so large and capable.

crisis if that proved necessary. Such use would probably not work against an established nuclear power such as Russia, because that power would presumably know that the United States had generated its nuclear forces and take appropriate countermeasures.[14] However, if the United States were generating nuclear forces for potential use in a regional conflict or for some special application, the need to generate nuclear forces might not matter. Actions could be more deliberate.

Evaluating specific dealerting proposals requires focused analysis. Details could matter a lot. For example, alerts do a number of things. One of them is to increase—usually substantially—the fraction of nuclear forces that could survive a first strike. By traditional strategic calculus, that increase should reduce incentives to strike first and, therefore, be considered stabilizing and "good." Alerts can, in various ways, decrease rather than increase the sensitivity and stability of nuclear force interactions during a crisis. Indeed, alerts have pluses and minuses.[15] Dealerting focuses only on the minuses, and specific schemes may not have the intended effects. Obviously, survivability of the forces is an issue, as is having a reasonable plan for regenerating the forces under difficult conditions should the need arise. Feiveson and his colleagues, for example, dismiss too cavalierly the problems of creating instabilities in crises by generating forces.[16] The act of force generation sends a signal, and political leaders may well be either too hesitant or too eager to generate forces as a result. This problem came up periodically during the Cold War, most recently with rail garrison MX.

It is premature to come to a conclusion on dealerting and deep cuts. Proponents may not have worked out all the details, but opponents have been even less convincing in their critiques. If all that the United States expects its nuclear forces to do is deter through threat of retaliation, there is probably some much smaller force, perhaps operated differently, that would suffice.

---

[14]Kahn (1969, pp. 268–269) discusses the dangers of a "mobilization race," and uses the analogy of World War I to illustrate the dangers that competitive force generation could cause in the nuclear age.

[15]See Buchan (1992).

[16]Feiveson (1999), pp. 121–122.

"Business as Usual, Only Smaller"

"Business as usual, only smaller" best characterizes the current official U.S. nuclear posture and strategy as described by Warner (1999) and easily inferred in the aftermath of the Nuclear Posture Review (NPR). Although the NPR was never made public, the general thrust was easy to discern from all the visible things that did not change. Subsequently, in her exquisite vivisection of the NPR, Nolan showed how it inevitably reinforced the status quo.[17] The basic U.S. force structure has remained unchanged, although it will shrink substantially now that the Russian Duma has ratified START II. It will shrink much more if START III negotiations proceed as planned.

The forces, the composition of the forces, and the operational procedures are more appropriate for the Cold War than the contemporary world scene. (About the best argument in their favor is the old saw, "If it ain't broke . . ."). Still, this approach cannot be sustained indefinitely. The force is larger than it needs to be if deterrence by threat of nuclear retaliation is the sole objective of U.S. nuclear strategy. Even a mildly expanded target base that included selected targets in emerging nuclear powers as well as chemical and biological weapons facilities in a larger set of countries would not necessarily require the sort of force that the United States plans to maintain. What the planned force appears best suited to provide beyond the needs of traditional deterrence is a *preemptive counterforce capability against Russia and China.* Otherwise, the numbers and the operating procedures simply do not add up. (Ironically, what maintaining the current U.S. approach to nuclear strategy and force planning will probably do in practice is accelerate the erosion of U.S. nuclear capability because staying on "autopilot" requires the least thought—and, therefore, the least action—of any other option.)

A More Aggressive Nuclear Posture

The United States could try to exploit its currently dominant nuclear position more aggressively. Beyond deterrence, large-scale counterforce, and selected use against CBW and other countries' nuclear facilities, the United States might choose to use its nuclear weapons

---

[17]See Nolan (1999a,b).

more aggressively to solve any problems that were both important enough to use nuclear weapons and difficult to solve any other way.

Interestingly, such a strategy need not require a large nuclear force. It does, however, require:

- Targeting flexibility comparable to that used for conventional weapons.
- Surveillance and targeting support roughly comparable to that needed for conventional weapons.
- Nuclear expertise on theater planning staffs.
- Suitable training for nuclear operators.
- Incorporating nuclear weapons in exercises both to gain operational experience and to find out what works and what does not.

In addition to making a more aggressive nuclear strategy actually feasible, these actions will also send a message to others that the United States is serious about maintaining a nuclear war-fighting capability and has the will to use the weapons if necessary.

For truly tactical applications, relatively short-range air-delivered nuclear weapons have some advantages (e.g., a short time of flight against movable targets). Alternatively, longer-range weapons could be equipped with in-flight updates, a possibility that has been considered for decades. Even adding the capability to recall the weapons and send them to a safe alternate site for destruction or recovery is a possibility. Other than that, current weapons are probably adequate, even if not optimal, as the last chapter suggested, for selected applications.

Our analysis also suggested that nuclear weapons are probably not necessary for most foreseeable tasks *if the United States procures adequate advanced conventional weapons and all of the C4ISR systems and targeting support necessary to make them effective.* Still, the future is inherently unpredictable. A force adequate to both deter by threat of punishment and deal with any emerging situation would not necessarily have to be big, but it would have to be flexible. It would also allow a coherent nuclear strategy, both actual and declaratory.

Nuclear Emphasis

Finally, there is the possibility of a much more radical nuclear force, one that is the main focus of U.S. military operations. This would be a significant departure from anything the United States currently plans. In essence, it would do something similar to what Russia says it wants to do: again supplement U.S. strategic forces with a larger arsenal of smaller nuclear weapons to be the mainstays of U.S. combat capability. Presumably, the rationales for entertaining such an approach could include

- the possibility of saving money compared to maintaining a large conventional force
- an opportunity to exploit U.S. nuclear expertise while it still exists
- a dramatic deterioration of the international scene.

Such an approach, while possible in principle, seems hard to justify based on the state of the world as it seems to be evolving and the effectiveness of modern conventional weapons. At the very least, it would almost certainly end any hope of limiting nuclear proliferation.

SOME ADDITIONAL COMMENTS ON DETERRENCE BY THREAT OF PUNISHMENT

In addition to the usual observations about deterrence by threat of punishment, there are some additional things to say about contemporary application of these ideas. As noted earlier, nobody can be sure what deters effectively. In practical terms, that could become even more complex in the future. For example, if the United States still believed it necessary to target Russia as a deterrent, what would it target?

- The Russian economy seems hardly worth targeting with nuclear weapons, considering what bad shape it is in already. (Perhaps a handful of military-oriented industrial facilities might warrant attack with nuclear weapons.)

- Similarly, Russian conventional forces have deteriorated to the point that they hardly warrant nuclear attack.[18]
- Leadership attacks have always been problematic—good idea or bad, straightforward or very difficult, depending on the details of the situation.

Of the traditional target categories, that leaves only nuclear counterforce targets. Threatening those is a war-fighting issue, not one of

---

[18]It is worth distinguishing between targeting conventional forces with nuclear weapons as part of a strategy of deterrence by threat of punishment and targeting elements of conventional forces to defeat them in battle. During the Cold War, there was a rationale for comprehensive nuclear targeting of Soviet nuclear forces that went roughly as follows: Soviet leaders place great value on their conventional forces as instruments of asserting and extending their power. Thus, threatening those forces with nuclear weapons might deter the Soviet leadership from bad behavior more effectively than targeting the civilian economy or military-industrial base. (Of course, there was no need to choose between them. The United States simply targeted both.) The point, though, is that this was an argument about coercing the Soviet leadership, not defeating Soviet forces in the field.

Targeting Russian conventional forces to offset a local conventional imbalance is more of a tactical war-fighting issue. The general point still applies: They are hardly worth the bother. Russian conventional forces are not completely inept. Indeed, during the Kosovo conflict, the Russians managed to move a battalion-sized force from Bosnia to Kosovo literally under NATO's collective noses and occupy Pristina airport before NATO could react. Closer to home, Russia's conventional forces have been flexing their muscles in the most recent round of fighting in Chechnya. This time the Russians appear to be winning. However, their "victory," if that is what it turns out to be, demonstrates more about the brutality of Russian conventional forces than their competence. Still, there could be conflicts near the Russian border in which the Russians would have conventional superiority and probably would win any local conflict absent outside intervention. There are at least four reasons why deterring Russian actions in such conflicts by threat of nuclear punishment of the sort that has traditionally characterized the SIOP would be inappropriate and unnecessary:

- The United States would have little stake in most such quarrels and would probably not feel the need to respond beyond some ritual condemnation of Russian "aggression."
- If the United States did care about the quarrel, the Russians would have to consider the consequences of even a Russian "victory" (e.g., sanctions of various sorts, undermining whatever remained of the U.S.-Russian relationship and the effect on areas of more importance to Russia, encouraging a military buildup by others).
- If the United States chose, it could probably field conventional forces capable of redressing whatever local unbalances existed.
- If the United States did feel the need to resort to nuclear weapons, "tactical" use against the local forces would probably be more appropriate than any of the traditional SIOP-like options.

deterrence by threat of punishment. Thus, not only might small nuclear forces be adequate as a deterrent, but also even selecting suitable targets for those to threaten a traditional enemy could be difficult. A senior Russian defense analyst remarked recently that the United States and Russia need to find a different way to relate to each other than through a nuclear suicide pact.[19] He is probably right.

Finally, there is the question of whether a national policy of deterrence by threat of punishment will continue to be politically sustainable in the future. The punishment inflicted on Yugoslavia by precision conventional bombing during the Kosovo campaign to coerce Serbian acquiescence to NATO's conditions for peace undoubtedly caused some Americans to have moral qualms about inflicting even that degree of pain on civilian populations. The sustained economic sanctions on Iraq appear to have done much more damage to Iraqi civilians than to Saddam Hussein's government. With that sort of recent experience, will the American public continue to support a policy that threatens vastly greater damage?

MAINTAINING A ROBUST NUCLEAR DETERRENT[20]

One of the traditional concerns of U.S. nuclear strategy that continues today is how to maintain a robust deterrent capability. During the Cold War, that process was relatively straightforward, but it is much less clear in the contemporary world what a "robust deterrent" even means.

Figure 5.1 illustrates some of the problems. In the past, hedges were mainly against changes in the threat to U.S. forces (e.g., a technical breakthrough in antisubmarine warfare [ASW]) or an unexpected failure in part of the force (e.g., a systemic reentry vehicle problem). Such problems are still possible, and Figure 5.2 outlines some of the usual responses. This sort of activity was an integral part of the

---

[19]Based on conversations in Moscow between RAND analysts and Russian defense specialists.

[20]Most of this subsection is based on RAND analysis for the Defense Threat Reduction Agency (DTRA). The authors thank their colleague Jim Quinlivan for helping to develop some of these ideas.

RAND *MR1231-5.1*

- A hedge? Against what?
  - Changes in the external world
  - Unexpected failures of critical elements of the deterrent force
  - Changes in policy and "requirements" (e.g., What does the United States expect its nuclear forces to achieve? How does it measure "success"?)
  - Changes in the internal environment
- Must be evaluated in a broad policy context
  - Overall strategic objectives
  - Other military forces (e.g., conventional forces, defenses)
  - Supporting systems (e.g., tactical warning, command and control)
  - Constraints (e.g., cost, political acceptability)

Figure 5.1—What Constitutes a "Robust" Nuclear Deterrent?

Soviet-U.S. arms competition during the Cold War and sometimes helped fuel the arms race.

However, in a more complex—albeit less threatening—world, the standard reactions might be irrelevant or even wrong, as Figure 5.3 suggests. If all the United States expects its nuclear forces to do is inflict damage to punish or coerce enemies, then the traditional approach is probably adequate. In fact, absent a threat to the survivability of U.S. nuclear forces and command and control systems comparable to that mounted by the Soviet Union during the Cold War, guaranteeing the robustness of the U.S. nuclear forces against external factors should not be very demanding.

On the other hand, if the United States has different primary objectives—countering nuclear proliferation, for example—the demands on its nuclear forces could be different and robustness criteria could change accordingly. Similarly, the external world may require changes in the way the United States views its needs for nuclear forces. Figure 5.4 shows examples of what these differences might mean.

RAND MR1231-5.2

Problem	Counters	Observations
Technical breakthrough/new class of system (e.g., ABM defense, ASW)	• Develop or maintain technical counter (e.g., maneuvering reentry vehicles to counter ABM) • Abandon affected class of system • Operational counters	• Can be costly • Time to react can be an issue • May not work • Other systems have weaknesses as well • May be inadequate or dangerous
Operational change by adversaries (e.g., forward-deployed SSBNs)	• Operational changes (e.g., rebase bombers, increased alert rate)	• Can be costly, can be difficult in practice • Some operational changes are dangerous
Concern about component/sub-system failure (e.g., failure of a warhead design)	• Avoid using common components and subsystems in different systems • Maintain at least two similar sub-system designs for each class of system • Increase reliability testing	• Tends to be expensive • Difficult to do for small forces • Expensive • Inventory may be inadequate • May violate treaties
Increase force size by adversaries	• Compensatory increase in force size	• Costly • Arms race danger
Shortage of critical materials (e.g., critical nuclear materials)	• Increase supply (e.g., build more suitable nuclear reactors)	• Expensive • May be dangerous • May be politically difficult

Issue: How to design hedges that are affordable, effective, safe, and politically acceptable?

Figure 5.2—Traditional Types of "Hedge" Responses to Maintain a Robust Force

What this means as a practical matter is that the United States needs to address with precision exactly what it expects its nuclear forces to accomplish in the future and what exactly that means for the characteristics of the forces and the way they are operated. For example, as we noted earlier, maintaining a nuclear force that is intended only to

RAND *MR1231-5.3*

- The United States may have different objectives (e.g., more emphasis on deterring or preventing nuclear proliferation than on maintaining a stable nuclear balance between two major players)
- The external world has changed
  - Even old players may have changed in "bad" ways (e.g., Russian factions that are "Soviet" thugs in sheep's clothing)
  - New players who may not understand the "rules of the game" in the same way (may also see us more clearly than we see ourselves)

New "rules" could require *opposite* solutions

Figure 5.3—Why the "School Solution" Might Be Inadequate . . . or Even Wrong

deter an enemy from some set of actions by threat of retaliation is relatively undemanding by modern U.S. standards.[21] It requires a nuclear force and an associated command and control network that are survivable against enemy threats, and are reliable, safe, and affordable to operate. Numbers probably do not matter much nor does the ability to respond swiftly. By contrast, nuclear counterforce operations require higher-quality, faster-acting weapons—both of which the United States already has—and a force size to be determined by the size and the nature of enemy forces. An expanded view of nuclear war-fighting—actual "tactical" application—is more demanding at almost every level (e.g., weapons characteristics, responsive command, control, and communications [C^3], platform-weapon matches). On the other hand, depending on the alternatives

[21]The deterrent still could prove ineffective, for reasons we have already discussed. However, it is unlikely that the deterrent quality of the force would be sensitive to the details of the U.S. nuclear force structure. Even if it were, it is unlikely that the United States would understand the workings of potential enemies' minds well enough to know what those sensitivities were and adjust its forces accordingly.

RAND *MR1231-5.4*

Characteristics	Cold War model	Unstable post–Cold War model
Robustness criteria	• Assured second-strike capability • Second strike stability	• High-confidence, first-strike counterforce capability (against proliferators)
Adversaries	• Acknowledged and accepted opponents (e.g., Soviet Union)	• Amorphous, potentially anonymous • "Old" enemies, perhaps with new faces • New players • Non-nations states • "Crazy players" with nothing to lose
Means	• "Strategic" nuclear systems	• Full continuum of possibilities – Defenses – Conventional forces – Special operations forces – Nuclear forces – Other
Rules	• Common understanding of "rules of the game" – Communicated and understood at a sophisticated level • "Dialog" routine ➜ Bureaucracy handles things in routinized way	• Dialog either not permitted or not reliable • Our past actions more likely to be remembered by enemies than by us ➜ No commonly accepted rules of the game
Implications	• Desire for stability dominates • Focus on ritual balance measures	• Actual nuclear use a real possibility • Posture looks like a move toward preemption

Figure 5.4—How Using the Wrong "World Model" Could Lead to Erroneous Robustness Criteria for Nuclear Forces

available, nuclear forces may not be either required or adequate to deal with a wide range of contemporary problems, as Figure 5.5 suggests.

FLEXIBLE USE OF NUCLEAR FORCES

Numerous studies since the end of the Cold War[22] have called for the traditional, rigid SIOP planning process to be replaced by a much more flexible approach analogous to what is currently used to de-

---

[22]See, for example, Feiveson (1999), Buchan (1994), and National Academy of Sciences (1997).

RAND *MR1231-5.5*

- Other alternatives include:
  - Improved conventional weapons
  - Special operations forces
  - Defenses

  Issue: How effective are these options likely to be?

- Implications for U.S. nuclear forces
  - *Perhaps* U.S. nuclear forces are neither necessary nor sufficient to deal with new problems posed by the post–Cold War world
  - Issue: Does the United States need nuclear weapons to be anything other than terror weapons?
    - If not, "robustness" issues are straightforward (e.g., Wohlstetter criteria)
    - If actual warfighting use remains possible, retaining an operationally robust nuclear force will be much more complex

Figure 5.5—But These Problems May Not Require Nuclear Responses

velop conventional targeting plans.[23] Doing so is a necessary condition for extracting any value from nuclear weapons other than "existential deterrence" (i.e., the hope that enemies will be deterred from overly provocative acts—whatever that actually means—by the mere existence of U.S. nuclear weapons), defeating enemy capability with preplanned preemptive counterforce attacks, and deterrence by punishment from preplanned coercive attacks.

There are several reasons why this is true:

- The kinds of relatively rigid nuclear plans that the U.S. emphasized in the past are likely to be irrelevant in the contemporary world. At most, a few preplanned nuclear options, such as strikes

[23]Decades ago, SIOP planning was the most complex war planning that the United States did because of the number of weapons involved, the complications introduced by nuclear weapons effects (e.g., fratricide), the demanding timelines of large-scale nuclear operations, and the limited computer capability available. Now, with nuclear forces much smaller and simpler and much greater computational capability readily available, nuclear planning *should* be much less demanding than either large-scale conventional planning or special operations planning.

against enemy nuclear forces, might be relevant to contemporary U.S. security problems in the extremely unlikely event that situations requiring such drastic action were to materialize.

- Predictable kinds of war-fighting plans (e.g., interdiction, targeting, critical time-urgent targets) require flexible planning and execution capability. Conventional forces increasingly tend to operate that way. Nuclear forces would have to as well.
- Absent the set-piece nuclear scenarios of the Cold War, the details of situations dire enough to warrant U.S. nuclear use are so unpredictable that preplanning is likely to be useless. Feiveson et al. said it best:[24]

> The circumstances in which the United States might seriously consider the use of nuclear weapons are so uncertain and unforeseeable that it makes little sense to focus on a handful of preplanned options.

Although United States Strategic Command (STRATCOM) has increased the flexibility of its planning process considerably compared to the Cold War days, the process is still relatively slow compared to regular theater planning. Whether it is good enough is an analytical issue that requires more detailed examination.

Improving the mechanics of the planning process is necessary but not sufficient for developing operational capability to employ nuclear forces flexibly and "tactically." At least the following are also required:

- Targeting support roughly comparable to that needed for conventional forces
- Suitable command and control
- Nuclear planning expertise at the theater level
- Adequate training for nuclear operators
- Adjustments to the weapon systems, if necessary.

---

[24]Feiveson (1999), p. 56.

The first point merely reflects the need to be able to find, identify, and locate suitable targets, particularly mobile targets such as armored forces or mobile missiles, well enough and in a timely enough way to target nuclear weapons against them. The second means adding sufficient command and control capability to bring the weapons to bear effectively. This was a shortcoming that would have seriously complicated, for example, any U.S. attempt to use nuclear weapons to counter an Iraqi invasion of Saudi Arabia during the early stages of the Gulf War had such an event occurred.[25]

Nuclear expertise on theater planning staffs is both critical and currently in short supply. Unlike the Cold War days when nuclear weapons were an integral part of U.S. war plans and theater planners, especially in Europe, were well-versed in nuclear matters, today nuclear-use questions apparently do not arise very often. As a result, nuclear expertise at the theater level has atrophied.[26] Although U.S. STRATCOM retains the capability to make weapons effectiveness and collateral damage calculations (and, indeed, make them routinely), the potential users of the weapons may not be knowledgeable enough to ask the right questions in a timely enough manner to develop workable plans in a crisis. The practical consequence is that in some future crisis, planners may not understand the capabilities and limitations of nuclear forces well enough to identify sensible nuclear options, if indeed there are any.

A related issue is training. Much of the training for nuclear use—particularly simulated missile launching—is routine and relatively

---

[25]This is typical of one of a very small set of generic scenarios where U.S. nuclear use might be a serious tactical option. First, serious U.S. interests were at stake. Second, substantial numbers of American lives were at risk during the early stages of Operation Desert Shield when large numbers of American troops were in the theater, but not enough quality weapons and delivery systems were there yet to mount an effective conventional defense. Third, as a result, adequate conventional options were not available. Fourth, because of the nature of the theater—large, relatively unpopulated areas—it might have been possible to use even strategic-size nuclear weapons without excessive collateral damage. Faced with the prospect of substantial damage to U.S. interests, large-scale loss of American lives, no other viable military options, and little time to act, a U.S. president might well want to know if there is a nuclear option available that could solve the problem. Even then, of course, he would have to weigh the short-term benefits against the long-term consequences, but absent the capability there would be no options to consider.

[26]This observation is based primarily on conversations with U.S. STRATCOM staff.

independent of context, but some is not. For example, the United States retains a modest number of nuclear bombs in Europe for use by tactical aircraft. Tactical delivery of nuclear bombs requires practice if pilots are to retain their proficiency. Absent that, any actual nuclear capability is more illusory than real. That may be acceptable in situations where the weapons are intended only to send a political message and nobody contemplates actually using them. However, whatever message goes out for external consumption, U.S. planners at least need to know the difference.

A more important and difficult aspect of training is integrating nuclear weapons into an overall campaign. A necessary condition for doing that is including nuclear use in exercises;[27] exercises are the closest thing to actual combat for gaining experience with both the planning and operational aspects of nuclear use in a larger context. *Exercises are probably also the best "laboratory experiments" to identify what operational concepts for the use of nuclear weapons actually make sense, if any.*[28]

In sum, the capability for flexible use of nuclear weapons is a sine qua non for dealing with unanticipated crises, arguably the most likely kind of situation that could lead to actual use of U.S. nuclear weapons. If only a small number of weapons were needed, targeting the weapons should be trivial. However, if the nuclear strikes were designed for military—as opposed to political—effect, incorporating even small strikes into large-scale campaign plans could require considerable coordination. Strikes involving larger numbers of weapons or more complex tactical situations (e.g., proximity to friendly forces, civilian collateral damage) require much more effort, support, and preparation, which would entail a serious commitment to maintaining nuclear expertise.

---

[27]We have been raising this issue for some years now. See, for example, Buchan (1994), pp. 42–43.

[28]Using nuclear weapons in exercises also sends signals to others, of course. For example, the Russians conducted exercises in the late 1990s involving nuclear use that were presumably designed to buttress the credibility of their new declaratory policy of relying more heavily on nuclear weapons to protect their borders (Pry, 1999, pp. 264–265; Gordon, 1999; *Nezavisimaya Gazeta*, 1999). Regardless of any desire on their part to send political signals to the rest of the world, they would have to do such exercises in any case to develop real operational capability.

CHARACTERISTICS OF NUCLEAR WEAPON SYSTEMS

The characteristics of the particular nuclear weapon systems that the United States maintains in its inventory could be at issue as well. If the United States were to opt for heavy reliance on nuclear weapons for tactical use, as the Russians say they might do now, it would almost certainly lead to an arsenal of more numerous, smaller nuclear warheads (assuming, of course, that the United States preserves the capability to develop and deploy new warheads). That is a path the United States could still choose as long as it had sufficient critical nuclear materials (i.e., plutonium, tritium, enriched uranium), its nuclear weapon design skills, and the ability to develop new warheads (assuming that existing designs would not suffice).[29] However, that would run strongly counter to current political trends in the United States and would almost certainly be viewed as inflammatory on the international scene.

More fundamentally, the need is not really there, or at least it should not be if the United States plans its conventional forces properly.[30] Even in the conventional world, our past studies have shown that large numbers of relatively small (e.g., 500 lb), accurate weapons were more effective than fewer, larger (e.g., 2000 lb) precision-guided weapons. Thus, making a case for new tactical nuclear weapons, even so-called "mini-nukes," would be difficult unless both the external world and U.S. fortunes change drastically.[31]

Finally, as the previous chapter suggested, our review of nuclear weapons effectiveness against selected classes of demanding targets showed that current U.S. warheads are not particularly bad, although

---

[29]Developing new nuclear warheads has traditionally required testing. The Comprehensive Test Ban Treaty (CTBT) would have forbidden such testing. However, the U.S. Senate failed to ratify the CTBT, so that question is moot. Relying on computer simulation in lieu of testing is a theoretical possibility but has never had sufficient credibility with weapon designers.

[30]Buchan (1994), p. 12. It is not reassuring, however, that the Air Force in particular has a track record of underinvesting in weapons as opposed to platforms, in spite of the mass of studies that have emphasized the relative importance of weapons.

[31]We have made this point in the past (see Buchan [1994], p. 65). The real issue is whether trying to recast a nuclear "sledgehammer" for use in roles requiring a stiletto is worth the trouble, particularly when conventional "stilettos" are getting sharper all the time.

smaller warheads would probably be more appropriate in many situations. Using large strategic warheads in tactical situations (e.g., against moving armor), of course, involves considerable potential for collateral damage. However, if a tactical situation were desperate enough to contemplate using nuclear weapons, perhaps collateral damage would be less of an issue.

Beyond the warheads themselves, there are questions about the delivery systems. For most applications, any of the current U.S. systems is likely to be adequate. The specific exception is attacking mobile targets—e.g., invasion forces or mobile missiles. Even assuming good initial targeting, flight time of the weapons could be an issue, particularly for long-range ballistic or cruise missiles. For this reason, relatively short-range aircraft-delivered weapons are likely to be the weapons of choice for attacking such targets. However, gravity bombs are appropriate weapons only in a relatively benign air defense environment. Short-range attack missile (SRAM)-class weapons would be particularly appropriate, but the United States no longer has SRAMs in the inventory and does not, as far as we know, plan to develop a replacement. Alternatively, long-range missiles could be modified to receive in-flight targeting updates to allow targeting flexibility. The United States has considered such technical options for decades but has never felt the need to pursue them. In sum, there are several ways to solve this problem, and solving it is probably important *if the United States wants to be able to threaten mobile targets with nuclear weapons as part of its overall security strategy.*

EXPLOITING ASYMMETRIES

A general thread that runs through the analysis of future U.S. security needs is the importance of asymmetries. Coping with future nuclear threats may require more than just deterrence, and deterrence might need more than just nuclear threats. For example, as we noted earlier, a nuclear strike might not be an appropriate response *even to a nuclear attack.* Depending on who is responsible, a rifle bullet might be a more appropriate response and a more effective deterrent. Similarly, U.S. nuclear weapons might be a suitable response or deterrent to nonnuclear threats that were important enough, and not just to other so-called "weapons of mass destruction." The point is that *U.S. security strategy in general needs to be richer and more nuanced to deal with contemporary security problems.* It needs to be able to distinguish between threats—nuclear or otherwise—that can best be deterred by fear of a nuclear response, those that can be most effectively and credibly deterred by other kinds of responses, and those that can be defeated in either a traditional military fashion or by other means. *U.S. nuclear strategy needs to be flexible enough to recognize and exploit these asymmetries.*

NUCLEAR PROLIFERATION

Another contentious issue affecting future U.S. nuclear strategy is the effect of U.S. policy on nuclear proliferation. Central to the arguments of many advocates of dramatic nuclear arms reductions and dealerting is the presumption that other nations will be more inclined to acquire nuclear weapons themselves unless the United States and other established nuclear powers drastically reduce or eliminate theirs. That presumption is codified in the NPT.

That assumption may be true, but the issue is not so clear cut. Maintaining high-quality U.S. nuclear forces might be a more effective deterrent to some states considering joining the nuclear club. In essence, the message to those states would be, "We may not be able to stop you from acquiring nuclear weapons, but if you do, you put yourselves at much greater risk if our interests ever collide. And don't expect us to bail you out if you get in trouble!" Thus, U.S. nuclear capabilities could either encourage or dissuade others from developing nuclear weapons. Individual cases could vary considerably.

Most likely, however, regional powers will make decisions about acquiring nuclear weapons with little regard to what the United States or others think or do themselves. Even if they are willing to be influenced by American views, countries are notoriously bad at communicating their positions effectively and reading each other accurately.[32] There will be the usual protests from the international community and probably some set of diplomatic and economic sanctions, as occurred in the case of India and Pakistan. Presumably,

---

[32] See, for example, Perkovich (1999, p. 421) for a description of how the United States and India misread each other prior to the most recent Indian test series.

proliferators will have anticipated such responses and already taken them into account in their decisionmaking processes. Then, as in the case of India and Pakistan, a prolonged bargaining process to manage each proliferation incident will probably occur. Also, as in the case of India and Pakistan, such bargaining processes may well fail.

Although the United States has focused most of its concerns about nuclear proliferation on so-called rogue states (since June 2000 known as "states of concern"—potential enemies such as North Korea, Libya, Iraq, and Iran) that are likely to develop modest arsenals at most, the possibility of major industrial powers, even those that are nominally U.S. allies, developing nuclear weapons might have a greater impact on the world in general and the United States in particular in the long run. For example, Japan, with its large supply of plutonium and world-class technical expertise, could very quickly become at least a major regional nuclear power if it chose to. That would probably immensely complicate the Asian security situation, affecting nuclear powers Russia, China, India, and Pakistan, whether directly or indirectly, and probably influencing nuclear decisions in North Korea, South Korea, and Taiwan. Inevitably, the alterations in the regional and global balances of power would affect the United States as well, even if Japan remained a United States ally. In fact, the net effect on the United States could be much greater than, say, a handful of North Korean nuclear weapons. Thus, Carl Builder (1991) might have been premature in pronouncing nuclear future weapons to be the weapons of the weak.

Finally, although the United States has made clear its opposition to further proliferation of nuclear weapons, long-range delivery systems, or other particularly threatening weapons, there are limits to how much the United States should be willing to pay to prevent proliferation. Otherwise, potential proliferators have too much bargaining leverage and, therefore, incentives to develop nuclear or other advanced weapons. That is a dangerous game. Ukraine, for example, played its hand skillfully on the issue of giving up its nuclear weapons when the Soviet Union dissolved. It managed to get some political and economic benefits in exchange for giving up its nuclear weapons without antagonizing Russia excessively or alienating the United States. North Korea has "pushed the envelope" in trying to extract the maximum political mileage from its latent nuclear capability. So far, it has worked, but the future remains uncertain and potentially very dangerous for North Korea and for others.

The point is that nuclear proliferation issues are complex. Individual countries' decisions about whether to develop nuclear weapons are probably going to be influenced only at the margin by U.S. nuclear strategy. Even then, it is not at all clear whether U.S. restraint or enhanced capability is more likely to discourage proliferation. The most effective approach is probably a balanced combination of the two: minimizing the importance of U.S. nuclear capabilities in most international dealings as a "carrot," but making it clear that the United States has nuclear and conventional weapon systems capable of dealing with an emerging nuclear power more forcefully should the need arise as either a spoken or unspoken "stick." That is basically the policy that the United States has followed for some time.

IS "WITHERING AWAY" INEVITABLE?

Earlier, we raised the possibility that U.S. nuclear capability might wither away over time. If that were to occur, then the United States would face elimination of its nuclear arsenal some time in the future whether it liked it or not, and U.S. security strategy would have to reflect that reality. This is a serious issue, and is increasingly recognized as such.[33]

Problems occur at several levels. First is the country's nuclear infrastructure. This issue has received a considerable amount of attention, primarily from the Department of Energy, which has the responsibility for maintaining the U.S. nuclear stockpile. The United States has shut down its production reactors that produce plutonium. It has drastically reduced its capability to produce key nuclear weapon components (e.g., "pits" for bombs). It relies heavily on its stockpile of existing warheads for both warhead components and nuclear materials. At the moment this problem appears to be under control, but there is not much slack.

More fundamental, and much harder to solve, are problems of retaining expertise in critical areas. For example, the United States

---

[33]See, for example, Defense Science Board (1998) and Buchan (1994).

leads the world in nuclear weapons design expertise. That expertise is going to be difficult to maintain at a time when the United States plans no new nuclear warheads. To be sure, nuclear weapons designers must maintain the current nuclear stockpile and implement the science-based stockpile stewardship program. With the U.S. Senate's failure to ratify the CTBT, the United States could conduct nuclear weapons tests again, although it currently has no plans to do so. If it did decide to test again, designing new warheads would be an option, although there currently appears little need to do so. The problem is that given the relatively low priority that nuclear weapons issues in the United States and the low likelihood that a weapons designer would get a chance to design new nuclear warheads, there are few career incentives for the "best and brightest" to get into this business. Thus, over time, U.S. capability will erode. Even if the science can be preserved, the art and engineering of warhead design are likely to be lost.

The same is true in other weapon system-related areas as well, although in some cases to a somewhat lesser extent. For example, absent plans to build new ICBMs or SLBMs, there is little incentive for the few aerospace companies that still have the design skills to build big missiles to maintain their capabilities. Similarly, there is little incentive for bright young engineers to choose this career path. Because the United States still needs space boosters, maintaining that capability should alleviate the missile design problem to some degree. Still, missile system integration for nuclear weapons application remains an issue, as do specific subsystem technology areas (e.g., reentry vehicle technology, where the United States has traditionally led the world).

Aircraft and air-delivered weapons are less of a problem, because the United States needs bombers and cruise missiles for conventional applications. Although nuclear-capable platforms generally require special hardening, that is not nearly as big an issue as the design of the platforms and delivery vehicles themselves.

At least as important as the technical skills are those of the military operators. Given current service priorities, nuclear weapons skills and experience are likely to lose the luster that they once had. Traditionally, both in the Air Force and the Navy, nuclear service has been considered an elite assignment and was sought after accordingly. With the current general lack of interest in nuclear issues, it will be difficult to persuade talented officers and enlisted personnel to enter nuclear career fields. Even including nuclear skills in the "tool kits" of officers on planning staffs will be difficult, as current experience suggests. In some cases—bomber operations, for example—there should be less of a problem, because nuclear operations are not that different from conventional operations. On the other hand, the most specialized skills associated with handling nuclear weapons are going to be harder to maintain.

These problems will be extraordinarily hard to solve because solutions will require influencing the decisions of large numbers of disparate individuals as well as various organizations, large and small, public and private. During the Cold War, it was easy to persuade individuals and organizations of the importance of dealing with nuclear weapons. Now, that will be more difficult. *Understanding the practical constraints that this problem imposes will be critical in shaping future U.S. nuclear strategy.*

Chapter Six

CONCLUSIONS

Nuclear weapons remain the ultimate guarantor of U.S. national security. Because of the massive destruction that even a single nuclear detonation could cause and the amount of explosive power that can be packed into a very small package, nuclear weapons trump all other types of weapons either as a deterrent—a threat of punishment—or as a military instrument to be used if the situation were serious enough to warrant such drastic action. Even when not actually used or overtly brandished, their mere existence in the U.S. arsenal provides implicit leverage in any serious crisis. They form a nuclear "umbrella" over all other U.S. military forces and instruments of policy.

However, nuclear weapons have significant disadvantages as well, most of which result from the same characteristics that make them potentially attractive:

- Their sheer destructiveness means that use of nuclear weapons, particularly on a large scale, is likely to produce damage out of proportion to any reasonable military or political objective. As a result, a *tradition of non-use has evolved, a tradition that particularly serves the interest of the United States.*
- Battlefield use of U.S. nuclear weapons can cause headaches for field commanders—e.g., radiation, blackout, fallout, problems obtaining release authority, planning problems. Such problems associated with actual employment of nuclear weapons may make their use more trouble than it is worth unless the need is overwhelming.

- Because the consequences are so great, the need for safeguards to avoid accidents, incidents, unauthorized use, mistakes, or theft of nuclear weapons is overwhelming. *The weight given to this factor in the equation will have a major impact on the future nuclear strategy that the United States selects and how it chooses to implement that strategy.* It is one of the two or three factors at the heart of the current dispute over future U.S. nuclear policy.

As a mature and experienced nuclear power—especially one that also dominates the conventional military and economic arenas—the United States has a variety of choices in crafting a nuclear strategy for the future. Also, even more than in the past, the United States has an overwhelming interest in preserving its current place in the world. It is both prosperous and secure, with no threat on the horizon even approaching that posed by the former Soviet Union. It needs, then, to design a national security strategy flexible enough to deal with the future however it evolves and shape that future to the degree possible.

Deciding where nuclear weapons fit is a central part of that process. Choosing an appropriate role for U.S. nuclear weapons will require balancing potentially competing objectives:

- Extracting the appropriate value from its nuclear forces (i.e., imposing its will on others in situations where it really matters).
- Making nuclear weapons *in general* less important rather than more important in world affairs to reduce the incentives for others to acquire them.
- Avoiding operational practices that might appear overly provocative to other nuclear powers and prompt unfortunate responses (e.g., reliance on launch-on-warning or preemption).
- Operating nuclear weapons in such a way that risks of accidents, unauthorized use, and theft are minimized.

There are several general nuclear strategies that the United States might adopt. Each has different implications for force structure and operational practice.

The most obvious transcendent role for U.S. nuclear weapons in the current world is to continue to provide a deterrent force capable of threatening any nation (or nonnation that controls territory or valuable facilities) with massive destruction. That is what nuclear weapons are particularly well-suited to do.

The political payoff from such a strategy could be problematic, however. All deterrence and coercion strategies suffer from the common weakness that they depend for success on decisions made by enemies. Empirically, it is extraordinarily difficult to be sure what deters whom from doing what to whom. Credibility is a key issue as well. Even if the United States means a threat seriously, others may not believe it and act accordingly. The United States would then be faced with the classic problem of needing options to act if deterrence should fail.

Still, the only real threat to the United States' existence as a functioning society remains Russia's nuclear arsenal, even if it shrinks to much lower levels, as projections suggest. Even with the chilling of U.S.-Russian relations since the post–Cold War "honeymoon" ended, it is unlikely that the Cold War nuclear standoff between the United States and Russia would return with the same force as in the old days. If it did, or if other similar threats emerged, the familiar solution of deterrence by threat of nuclear retaliation, with all its theoretical flaws, is still probably the best option for the foreseeable future. In the contemporary world that probably requires:

- Survivable forces and command and control, as in the past.
- A force of almost any reasonable size. (Damage requirements were always largely arbitrary. In the contemporary world, there is an even less compelling need for a large force. For example, if the United States were to target Russia, what would it target? The economy and the conventional military hardly seem worth attacking with nuclear weapons. Attacking leadership is problematical. That leaves only strategic forces, and targeting them is a separate strategic issue. *It would be a supreme irony of the contemporary world if strategic forces were the only suitable Russian targets for U.S. nuclear weapons now when such attacks would have been ineffective and possibly counterproductive during the Cold War.)*

- An adequate mix of forces to hedge against technical or operational failures: *The key Air Force systems to ensure variety are air-breathing weapons* (i.e., bombers and cruise missiles).

An important point is that there is no need for a prompt attack. Indeed, prompt responses could be dangerous under some conditions. That means that even small, dealerted forces could, in principle, have considerable deterrent power if they solved the practical problems (e.g., survivability, force generation) adequately.

These are familiar problems from the old Cold War days with some modifications to accommodate the changes in the relationship between the United States and Russia. The contemporary world has some new wrinkles in addition to the usual elements:

- Identifying attackers may be harder with more players and diverse delivery options available.
- A broader range of options than just nuclear weapons may be needed to deter or deal with some kinds of threats (e.g., terrorists who cannot be threatened directly by U.S. nuclear weapons).
- No threat of punishment may be sufficient to deter some nuclear threats to the United States (e.g., nations with nuclear weapons and nothing left to lose). *An established nuclear power coming unglued and lashing out is the worst possible threat to U.S. security for the foreseeable future,* much worse than so-called rogue nations. Something other than deterrence will be necessary to deal with them.

A more challenging issue is the degree to which the United States wants to include actual war-fighting use of nuclear weapons in its overall strategy. The first possibility is nuclear counterforce. Ironically, nuclear counterforce, which probably would not have worked during the Cold War, might be feasible in the current world, particularly against new nuclear powers that have not learned how to play the game (i.e., have not developed high-quality mobile systems and survivable command and control). A counterforce emphasis would provide a more quantitative basis for sizing forces than "simple" deterrence. It would also put more of a premium on timely delivery. Also, to the degree that U.S. nuclear strategy included counterforce as a hedge against nuclear proliferation, it could be viewed as part of the "robustness" criteria normally associated with keeping a deterrent force effective (e.g., multiple types of systems, different key components, etc.).

The current U.S. counterforce advantage is probably fleeting. Counters are well known. They just require resources, time, and experience to implement. Thus, there is an issue about how much contemporary U.S. nuclear strategy ought to emphasize counterforce. To some degree, the strategic issue is almost moot; any nuclear force the United States maintains is likely to have considerable inherent counterforce capability if it operates more or less the way U.S. strategic forces operate currently. Interestingly, *only a large-scale commitment to a counterforce-heavy strategic doctrine is likely to require the "business as usual, only smaller" type of force structure recommended by the original NPR.* That point will not be lost on others who infer U.S. intentions from the force structure that they observe and might react badly to what they could view as a serious U.S. threat. They will probably not be much impressed by "bureaucratic momentum" as an explanation for the United States maintaining large nuclear forces structured as they were during the Cold War.

Using nuclear weapons against a broader set of military targets is a policy option as well. It is a more interesting possibility because it follows a broader policy logic: *One of the reasons the United States maintains nuclear weapons is to deal with any situation that should emerge that threatens vital U.S. interests and cannot be dealt with adequately in any other manner.* As we suspected, the real issue is conventional weapons effectiveness. If the United States invests adequately in advanced conventional weapons, there should be no need for nuclear weapons to be used "tactically" except for attacks on deeply buried targets if that proved to be necessary. The only potential exception is a conflict against a world-class enemy fought at long range where even effective conventional weapons did not provide sufficient mass of firepower to solve the tactical problems at hand. In such a case, large numbers of small U.S. nuclear weapons might provide the added firepower to tip the balance. Failing that, however, or a U.S. decision not to buy sufficient advanced conventional firepower, nuclear weapons are unnecessary and probably inappropriate for most tactical operations in which the United States is likely to become involved. Thus, *decisions on future U.S. nuclear strategy depend critically on issues not associated directly with nuclear weapons* (e.g., conventional weapons, ballistic missile defense).

If the United States wanted to maintain the option to use nuclear weapons tactically if a really desperate need arose, the problems are not generally with the weapons themselves but in planning and operational flexibility.[1] Such flexibility is the sine qua non for adapting to unforeseen circumstances. Indeed, there is a strong a priori case for developing this kind of operational flexibility for U.S. nuclear forces precisely because the circumstances under which U.S. nuclear weapons might actually have to be used in the future are so hard to predict that they cannot be planned for in advance.

Achieving such nuclear operational flexibility would require radical changes in U.S. nuclear operational practice. It would require at the very least:

- Suitable planning systems (e.g., near real-time target planning)
- Training
- Including nuclear weapons in exercises
- Nuclear expertise on theater planning staffs
- Suitable command and control
- Intelligence support comparable to that needed by conventional forces.

In the long term, there are other practical problems to solve if the United States is to remain a viable nuclear power. "Withering away" of U.S. nuclear operational expertise, support infrastructure, and

---

[1]However, some tactical applications appear to favor air-delivered weapons, particularly relatively short-range weapons. See Buchan (1994).

There is an extreme version of this argument that would call for a large number of very small nuclear weapons ("mini-nukes"). Our analysis suggested that such an option would be difficult to support. In fact, our previous work has shown that most large-scale conflicts could be best handled with large numbers of small (e.g., 500 lb) accurate conventional weapons and only a modest number of larger (e.g., 1000–2000 lb) conventional weapons. (See Buchan et al., 1994 and Frelinger et al., 1994.) Thus, even "mini-nukes" would be overkill for most applications. Still, if the United States were to take tactical use of nuclear weapons seriously, a larger force of smaller warheads would be more appropriate.

weapons-design capability may be unavoidable, given current career incentives, fiscal constraints, political realities, and service priorities. Thus, U.S. nuclear capability may diminish over time whether it likes it or not.

In considering overall contemporary U.S. strategic options, one striking possibility is that a new strategy could simultaneously be both more "dovish" and more "hawkish." That might involve a much smaller nuclear force intended to deter egregious behavior with threats of retaliation, but operated flexibly enough so that the weapons could in fact be used if a serious enough need arose against whatever particular set of targets turned out to be important. That sort of nuclear strategy would lend itself to a succinct description along the following lines:

"The United States views nuclear weapons as the ultimate guarantor of its security. They provide a means for deterring an enemy from damaging vital U.S. interests by threatening to punish him with massive damage. In particular situations, the United States might use nuclear weapons directly to resolve a crisis if vital U.S. interests were at stake and other means appeared inadequate."

Such a nuclear strategy would have to be supplemented by a broader spectrum of options to deal with contemporary problems that nuclear threats or use alone could not handle. In addition, working out the appropriate nuclear force structure to implement whatever strategy the United States chooses will require more detailed analysis. Ironically, force structure issues are likely to turn on relatively mundane issues, such as where the "knees" in the cost curves turn out to be. That, in turn, could affect the U.S. choice of a grand strategy.

It is a virtual certainty that any overall nuclear strategy the United States chooses will require a substantially different set of nuclear forces and operational practices than it has at present. *Proving that it can overcome the massive momentum that has shaped its past nuclear strategy and force structure decisions will be a major hurdle that the U.S. nuclear bureaucracy will have to clear in moving toward a sensible future nuclear policy. The range of possible policy options needs to be evaluated in much more detail than it has to date for the United States to choose a sensible nuclear strategy for the future.*

REFERENCES

AFX News, "Chinese Officials Threaten Nuclear Attack on U.S. Over Taiwan: Winston Lord," March 18, 1996.

Allison, Graham, and Phillip Zelikow, *Essence of Decision: Explaining the Cuban Missile Crisis*, 2nd ed., New York: Addison Wesley Longman, Inc., 1999.

Arkin, W., R. Norris, and J. Handler, *Taking Stock: Worldwide Nuclear Deployments 1998*, New York: Natural Resources Defense Council, March 1998, Appendix A and Table 2. The report can also be found at: http://www.nrdc.org/nrdc/nrdcpro/tkstock/download.html.

Associated Press, "France Sets Off Fifth Nuclear Test in the Pacific," *New York Times*, December 28, 1995, p. 13.

Blair, Bruce G., *Global Zero Alert for Nuclear Forces*, Washington, D.C.: Brookings Institution, 1995.

________, *The Logic of Accidental Nuclear War*, Washington, D.C.: Brookings Institution, 1993.

________, *Strategic Command and Control: Redefining the Nuclear Threat*, Washington, D.C.: Brookings Institution, 1985.

________, Harold A. Feiveson, and Frank N. von Hipple, "Taking Nuclear Weapons Off Hair-Trigger Alert," *Scientific American*, November 1997, pp. 74–81.

Bracken, Paul, *The Command and Control of Nuclear Forces*, New Haven, Connecticut: Yale University Press, 1983.

Brodie, Bernard, *Strategy in the Missile Age*, Princeton, New Jersey: Princeton University Press, 1959.

________ (ed.), *The Absolute Weapon: Atomic Power and World Order*, Chestnut Hill, Massachusetts: Harcourt, Brace, 1946.

________, *Escalation and the Nuclear Option*, Princeton, New Jersey: Princeton University Press, 1966.

Buchan, Glenn C., *Nuclear Weapons and the Future of Air Power*, RAND (forthcoming).

________, *One-and-a-Half Cheers for the Revolution in Military Affairs*, RAND, P-8015-AF, October 1997.

________, "De-Escalatory Confidence-Building Measures and U.S. Nuclear Operations," in Joseph L. Nation (ed.), *The De-escalation of Nuclear Crises*, New York: St. Martin's Press, 1992.

________, *U.S. Nuclear Strategy for the Post-Cold War Era*, RAND, MR-420-RC, February 1994.

________, et al., *Future Bomber Force Study*, RAND, R-4183-AF, August 1994. Government publication; not releasable to the general public.

________, *Employment of B-52s in Conventional Operations*, RAND, N-3628-AF, August 1994. Government publication; not releasable to the general public.

________, David Frelinger, and Tom Herbert, *Use of Long-Range Bombers to Counter Armored Invasions*, RAND, WP-103, March 1993.

Builder, Carl H., *The Future of Nuclear Deterrence*, RAND, P-7702, 1991.

Bundy, McGeorge, William J. Crowe, Jr., and Sidney D. Drell, *Reducing the Nuclear Danger: The Road Away from the Brink*, New York: Council on Foreign Relations, 1993.

Bundy, McGeorge, *Danger and Survival: Choices About the Bomb in the First Fifty Years,* New York: Random House, 1988.

Butler, Lee, "A Voice of Reason," *Bulletin of the Atomic Scientists,* May/June 1998, pp. 58–61.

Craig, Campbell, *Destroying the Village: Eisenhower and Thermonuclear War,* New York: Columbia University Press, 1998.

Chopra, V.D., and Rakesh Gupta, *Nuclear Bomb and Pakistan: External and Internal Factors,* New Delhi: Patriot Publishers, 1986.

Cohen, Avner, *Israel and the Bomb,* New York: Columbia University Press, 1988.

Defense Intelligence Agency, *Physical Vulnerability Handbook for Nuclear Weapons,* OGA-2800-23-92. January 1992. Government publication; not releasable to the general public.

Defense Science Board, *Report of the Defense Science Board Task Force on Nuclear Deterrence,* Washington, D.C.: Office of the Under Secretary of Defense for Acquisition and Technology, October 1998.

Douhet, Giulio, *The Command of the Air,* New York: Coward-McCann, 1942.

Efron, Sonni, "'Missile Attack' on Russia Was Just a Science Probe," *Los Angeles Times,* January 26, 1995, p. 1.

Ehrlich, Representative Robert, Hearing of the National Security, International Affairs and Criminal Justice Subcommittee of the House Government Reform and Oversight Committee, *Federal News Service,* May 30, 1996.

Feiveson, Harold A. (ed.), *The Nuclear Turning Point: A Blueprint for Deep Cuts and De-Alerting of Nuclear Weapons,* Washington, D.C.: Brookings Institution Press, 1999.

Federation of American Scientists, http//www.fas.org/nuke/guide/pakistan/nuke/chron.htm, "Pakistan Nuclear Weapons–A Chronology," Update June 3, 1998.

Freedman, Lawrence, *The Evolution of Nuclear Strategy*, New York: St. Martin's Press, 1986.

Frelinger, David, et al., *Use of Heavy Bombers in Conventional Operations*, RAND, N-3588-AF, August 1994. Government publication; not releasable to the general public.

Friedman, Norman, *The Fifty Year War: Conflict and Strategy in the Cold War*, Annapolis, Maryland: Naval Institute Press, 2000.

Friedman, Thomas L, "The French Ostrich," *The New York Times*, October 4, 1995, p. 21.

Gertz, Bill, "Russians Practiced Nuclear Counterattack on NATO," *The Washington Times*, July 8, 1997, p. A1.

________, *Bomber Flexibility Study: A Progress Report*, RAND, DB-109-AF, 1994.

Glasstone, S., and P. Dolan, *The Effects of Nuclear Weapons*, Washington, D.C.: U.S. Department of Defense and U.S. Department of Energy, 1977.

Gordon, Michael R., "Maneuvers Show Russian Reliance on Nuclear Arms," *The New York Times*, July 10, 1999, p. A1.

Herken, Gregg, *Counsels of War*, New York: Knopf, 1985.

________, *The Winning Weapon, The Atomic Bomb in the Cold War: 1945–1950*, Princeton, New Jersey: Princeton University Press, 1988.

Hersh, Seymour M., *The Samson Option: Israel's Nuclear Arsenal and American Foreign Policy*, New York: Random House, 1991.

Holloway, David, *Stalin and the Bomb*, New Haven, Connecticut: Yale University Press, 1994.

Kahn, Herman, *On Thermonuclear War*, 2nd ed., New York: Free Press, 1969.

________, *Thinking About the Unthinkable*, New York: Avon, 1968.

________, *Thinking About the Unthinkable in the 1980s*, New York: Simon and Schuster, 1984.

________, *On Escalation: Metaphors and Scenarios*, New York: Praeger, 1965.

Kaplan, Fred, *The Wizards of Armageddon*, New York: Simon and Schuster, 1983.

Kempster, Norman, and Doyle McManus, "Huge Warhead Cuts Approved: Bush, Yeltsin Act to End the 'Nuclear Nightmare,'" *Los Angeles Times*, June 17, 1992, pp. A1, A6, and A8.

Lewis, John Wilson, and Xue Litai, *China Builds the Bomb*, Stanford: Stanford University Press, 1988.

Marshall, Tyler, and Jim Mann, "Goodwill Toward the U.S. Is Dwindling Globally," *Los Angeles Times*, March 26, 2000, pp. A1, A30–A31.

Matonick, David, and Calvin Shipbaugh, *Selected Comparisons of Nuclear and Conventional Weapons Performance*, RAND (forthcoming). Government publication; not releasable to the general public.

McManus, Doyle, "U.S. Casts About for Anchor in Water of Post–Cold War World," *Los Angeles Times*, March 27, 2000, pp. A6, A8.

McNamara, Robert S., "The 'No-Cities' Doctrine," in Robert J. Art and Kenneth N. Waltz (eds.), *The Use of Force: Military Power and International Politics, Third Edition*, Lanham, Maryland: University Press of America, 1983.

Mohan, C. Raja, "Sundarji's Nuclear Doctrine," *The Hindu*, February 11, 1999. (http://www.ipcs.org/archives/a-ndi-index/99-02-feb.htm)

National Academy of Sciences, Committee on International Security and Arms Control, *The Future of U.S. Nuclear Weapons Policy*, Washington, D.C.: National Academy Press, 1997.

Nezavisimaya Gazeta, "Rossiyskaya armiya gotovitsya k otrazheniyu agressii," June 23, 1999. (http://www.armscontrol.org/ACT/March00/cwmroo.htm)

Nolan, Janne E., *An Elusive Consensus: Nuclear Weapons and American Security After the Cold War*, Washington, D.C.: Brookings Institution Press, 1999a.

Nolan, Janne E., "The Next Nuclear Posture Review," in Harold A. Feiveson (ed.), *The Nuclear Turning Point: A Blueprint for Deep Cuts and De-Alerting of Nuclear Weapons*, Washington, D.C.: Brookings Institution Press, 1999b.

Nordon, Paul, "Hardness and Survivability Requirements," in Wiley J. Larson and James R. Wertz (eds.), *Space Mission Analysis and Design*, 2nd ed., Columbia, Maryland: Microcosm, Inc., 1996.

Paddock, Richard C., "Lebed Says Russia Lost Track of 100 Nuclear Bombs," *Los Angeles Times*, September 9, 1997, p. A4.

Pape, Robert A., *Bombing to Win: Air Power and Coercion in War*, Ithaca, New York: Cornell University Press, 1996.

Payne, Keith B., *Deterrence in the Second Nuclear Age*, Lexington: University Press of Kentucky, 1996.

Perkovich, George, *India's Nuclear Bomb*, Berkeley: University of California Press, 1999.

Pomfret, John, "China Ponders New Rules of 'Unrestricted War,'" *The Washington Post*, August 8, 1999, p. 1.

Pry, Peter Vincent, *War Score: Russian and America on the Nuclear Brink*, Westport, Connecticut: Praeger, 1999.

Quinlivan, James T., and Glenn C. Buchan, *Theory and Practice: Nuclear Deterrents and Nuclear Actors*, RAND, P-7902, 1995.

Rhodes, Richard, *Dark Sun: The Making of the Hydrogen Bomb*, New York: Simon and Schuster, 1995.

________, *The Making of the Atomic Bomb*, New York: Simon and Schuster, 1986.

Roman, Peter J., *Eisenhower and the Missile Gap*, Ithaca, New York: Cornell University Press, 1995.

Rosenberg, David Alan, "The Origins of Overkill: Nuclear Weapons and American Strategy, 1945–1960," *International Security*, Vol. 7, No. 4, Spring 1983, pp. 3–71.

Rosenthal, Andrew, "U.S. to Give Up Short-Range Nuclear Arms: Bush Seeks Soviet Cuts and Further Talks," *The New York Times*, September 28, 1991, pp. 1, 4, and 5.

Sagan, Scott D., and Kenneth N. Waltz, *The Spread of Nuclear Weapons: A Debate*, New York: Norton, 1995.

Sakharov, Andrei, *Memoirs*, New York: Knopf, 1990.

Schelling, Thomas C., *Arms and Influence*, New Haven, Connecticut: Yale University Press, 1966.

Smith, R. Jeffrey, "The Dissenter," *Washington Post Magazine*, December 7, 1997, pp. 18–21; 38–45.

________, *The Strategy of Conflict*, New York: Oxford University Press, 1963.

Trachtenberg, Marc, *History and Strategy*, Princeton, New Jersey: Princeton University Press, 1991.

Turnbull, Peter C.B., et al., *Guidelines for the Surveillance and Control of Anthrax in Humans and Animals*, 3rd ed., Geneva: World Health Organization, WHO/EMC/ZDI./98.6, 1998.

Turner, Stansfield, *Caging the Nuclear Genie: An American Challenge for Global Security*, Boulder, Colorado: Westview Press, 1997.

United States Air Force, *New World Vistas*, "Munitions," 1995, p. 26.

Warner, Edward L. III, "Statement of the Honorable Edward L. Warner III, Assistant Secretary of Defense for Strategy and Threat Reduction, Before the Senate Armed Services Subcommittee on Strategic Forces Hearing on Nuclear Deterrence April 14, 1999," Washington, D.C.: Federal Document Clearing House, Inc., 1999.

Werrell, Kenneth P., *Blankets of Fire: U.S. Bombers over Japan During World War II*, Washington, D.C.: Smithsonian Institution, 1996.

Weigel, Fritz, "Uranium" in *McGraw-Hill Encyclopedia of Science & Technology,* 7th Edition, McGraw-Hill Publishing company, New York, 1992.

Wohlstetter, Albert J., "The Delicate Balance of Terror," *Foreign Affairs,* January 1959, pp. 211–234.